SOLAR HEATING IN COLD REGIONS

AF216067

Solar Heating in Cold Regions

A technical guide to developing country applications

JEAN-FRANCOIS ROZIS
AND
ALAIN GUINEBAULT

Illustrated by Vahram Varjabedian

Practical
ACTION
PUBLISHING

Practical Action Publishing Ltd
25 Albert Street, Rugby,
Warwickshire, CV21 2SD, UK
www.practicalactionpublishing.com

French edition published by
Groupe de Recherche et d'Echanges Technologiques
213, rue la Fayette, 75010 Paris, France

© Intermediate Technology Publications Ltd, 1996

First published in 1996
Transferred to digital printing in 2008

The author has asserted their right under the Copyright, Designs and Patents Act 1988 to be
identified as author of this work.

All rights reserved. No part of this publication may be reprinted or reproduced or utilized in any
form or by any digital, electronic, mechanical, or other means, now known or hereafter invented,
including photocopying and recording, or in any information storage or retrieval system,
without the written permission of the publishers.

Product or corporate names may be trademarks or registered trademarks, and are used only for
identification and explanation without intent to infringe.

A catalogue record for this book is available from the British Library & Library of Congress

ISBN 978-1-85339-329-7 Paperback
ISBN 978-1-78044-597-7 Digital book

Citation: Rozis, J. (1996) *Solar Heating in Cold Regions; A technical guide to developing country
applications*, Rugby, UK: Practical Action Publishing https://doi.org/10.3362/9781780445977

Since 1974, Practical Action Publishing has published and disseminated books and information
in support of international development work throughout the world. All print editions are
produced and distributed via ethical and sustainable print on demand global facilities.

Practical Action Publishing is a trading name of Practical Action Publishing Ltd (Company Reg.
No. 01159018 | VAT 880 9924 76). All profits are covenanted back to its parent group, Practical
Action (Charity Reg. No. 247257).

The views and opinions in this publication are those of the author and do not represent those of
Practical Action Publishing Ltd or its parent charity Practical Action. Reasonable efforts have
been made to publish reliable data and information, but the author and publisher cannot assume
responsibility for the validity of all materials or for the consequences of their use.

Typeset by J&L Composition Ltd, Filey, North Yorkshire, UK

The manufacturer's authorised representative in the EU for product safety is Lightning Source
France, 1 Av. Johannes Gutenberg, 78310 Maurepas, France. compliance@lightningsource.fr

Contents

Foreword

This book is written mainly for technicians, architects and designers who are interested in solar heating systems in cold regions of developing countries.

Over the last decade GERES has been supporting a programme of dissemination of solar energy installations in Ladakh in the Indian Himalayas. Low-maintenance equipment, suited to locally available materials and local technical skills, has not only been successful in meeting household energy needs but has also led to the introduction of new income-generating activities for agriculture.

In Part One the cold regions are described with case studies, effectively to define the particular characteristics of these regions. Part Two describes typical installations, how to produce and install them, together with rules for sizing them. Part Three tackles the underlying theory.

This book is above all a technical guide to the design and production of solar installations in regions where heating is an issue of the utmost importance. A simplified method of calculating the thermal behaviour of solar-heated spaces is presented in detail. For those wishing to take this further, we are more than happy to provide more detailed information on other manual or computer simulation methods.

There has been insufficient exchange of experience and information on this subject in the past. We hope that this work will result in improved exchange of information and help to end the marginalization of solar technology activities.

Acknowledgements

Jean-Francois Rozis of Groupe Energies Renouvables et Environnement (GERES) devised and produced this book with the help of Alain Guinebault, as part of the evaluation of technological equipment carried out by GERES and Runamaqui in development projects in India and Peru.

This work was made possible thanks to help from:

- The European Commission,
- The Dutch Ministry of Co-operation and Development,
- The development department of the French Ministry of Foreign Affairs – individual and community initiatives.

Furthermore we wish to thank the following for their various contributions:

- The Ladakh Ecological Development Group,
- The Leh Nutrition Project (Ladakh),
- The TARA Association/Development Alternatives of New Delhi,
- Craterre Latin America (Sykvua Natuk and Alain Hays) in Grenoble,
- J.L. Izard of the School of Architecture in Marseilles,
- Robert Célaire and Paul Mirmont who initiated the development projects in Ladakh,
- The Runamaqui Association and Dominique Gobin for all information provided and for his editorial contribution,
- Maighread Holland for translating the book into English,
- The Communication and Exchanges department of the Groupe de Recherche et d'Echanges Technologiques (GRET) for contributing to re-reading of the document.

And in particular Vahram Varjabedian for the quality and originality of his illustrations as well as other members of the GERES team for support of various kinds.

PART ONE

THE ISSUES FOR SOLAR HEATING IN COLD REGIONS OF DEVELOPING COUNTRIES

1 What can solar energy offer?

SOLAR ENERGY appears to offer an attractive solution to the energy problems of developing countries: the intensity of solar energy is often high, its exploitation technically simple and decentralised in nature. The use of solar energy in developing countries should therefore make it possible to improve people's living conditions and it is clear that as an energy substitute for fossil fuel it can also help to reduce greenhouse gas emissions.

What does solar energy currently contribute and what could it contribute in the near future to the advancement of people in developing countries? A brief presentation of the various applications can provide a partial response.

Three main areas of solar energy use can be identified and they correspond to the three ways of converting solar energy:

o *Solar energy converted to biomass by photosynthesis*
This process, the basis of all life on earth, is a fundamental part of the energy picture. Biomass is an extremely important part of energy consumption in poor countries, particularly for cooking food and heating buildings. In some African countries wood energy represents up to 90 per cent of total national energy production.

o *Solar energy converted to electricity (photovoltaics)*
This method of conversion, discovered in 1954, was initially used in space research. Today it provides a particularly appropriate solution for the electrification of the most remote regions for a number of reasons: it is economically competitive with rural electrification from the grid, it is relatively simple to install and it is a technology which can be mastered in the most advanced developing countries. It is also hoped that electricity in the villages will help to strengthen a population which is strongly tempted to migrate to the towns.

o *Solar energy converted to heat*
This mode of conversion has long given rise to great hope and interest. The first concrete applications for solar drying and water desalination have thus been used for several centuries. Examples include a wooden solar still built in 1870 by a British engineer for a mining development in

3

Chile at an altitude of 1400 metres. With a surface area of 4460m^2, it produced 20m^3 of fresh water per day for over 40 years.

An application unique to solar energy: solar heating

Among the best-known and most relevant applications of solar thermal energy in developing countries are:

o domestic water heating
o cooking
o drying of agricultural products
o water desalination
o sterilization of clinical instruments
o refrigeration
o heating of domestic, agricultural and tertiary buildings.

For most of these applications it is considered that there are now sufficient research results, both for theoretical aspects (modelling of physical phenomena) and technical aspects (choice of materials, behaviour over time etc.). But there are many difficulties facing those who wish to take solar energy into account in development. The dissemination of solar cookers is a good example.

In fact, in spite of the major issues of desertification from excessive extraction of biomass and the very laborious nature of firewood collection, something which is mostly undertaken by women and children, solar cooking has never become widely accepted. It has not yet been possible to provide cooks with a solar cooker which is compatible with traditional food preparation practices. Indeed the disappointments of solar energy are as great as the hopes it engendered.

Although solar energy may make quite a small contribution to the total world energy balance today, we wanted to present in this book a collection of solar installations which are particularly suited to the cold regions of developing countries.

They provide part of the solution to:

o *Rural development*
  Solar greenhouses and hen houses for agriculture, when there is good access to markets, are very viable development tools in economic terms. For home consumption they serve to improve nutrition, particularly through vegetable production in winter.

o *Improvements in living conditions*
  Solar heating of homes, schools, hotels, hospitals, maternity clinics is among the many applications much appreciated where used. In fact it is often the only source of heating in rural areas where people hesitate to use space heating because of the cost or lack of fuel. Using solar energy in urban areas, where conventional heating systems are more widely

used, helps to reduce significantly greenhouse gas emissions (particularly carbon oxides).

In Part One we present the various factors to be grasped in the context of using solar energy for heating in cold regions.

o We will therefore start by defining the characteristics of a cold region;
o We will then ask why solar heating is suited to the cold regions of developing countries;
o We will then describe four experiments and applications which in each case demonstrate the major issues for solar energy usage;
o Finally, in Chapter 2, we will present a methodology for action which can assist the successful diffusion of solar energy in developing countries.

The cold regions of developing countries

Generally we associate developing countries with hot climates, thus forgetting that part of the poor population lives in cold areas. In fact, the area covered by cold regions in developing countries is as large as that of hot regions. It mainly comprises three large areas: the Andean Cordillera, the Himalayan chain and the Chinese plateau. The cold rural area of north-west China, which is included as a developing country according to the poverty threshold criterion, alone measures one and a half million square kilometres.

How do we define a cold region of a developing country?
There are several criteria for defining cold regions, arising from two approaches: climatological and socio-economic.

The climatological approach

The first parameter which characterizes a climate is its temperature level. The temperature at sea level varies considerably according to latitude, going from hot at the equator to cold near the poles. Climatologists have therefore defined three large climatic zones covering the earth's surface:

o *The intertropical zone*
This is limited on either side of the equator by *isotherms* – north and south – of the coldest month of the year equal to 18°C (average temperature of the month.)

o *The temperate zones*
These are the two zones outside the intertropical zone out to the isotherm of the warmest month equal to 10°C.

5

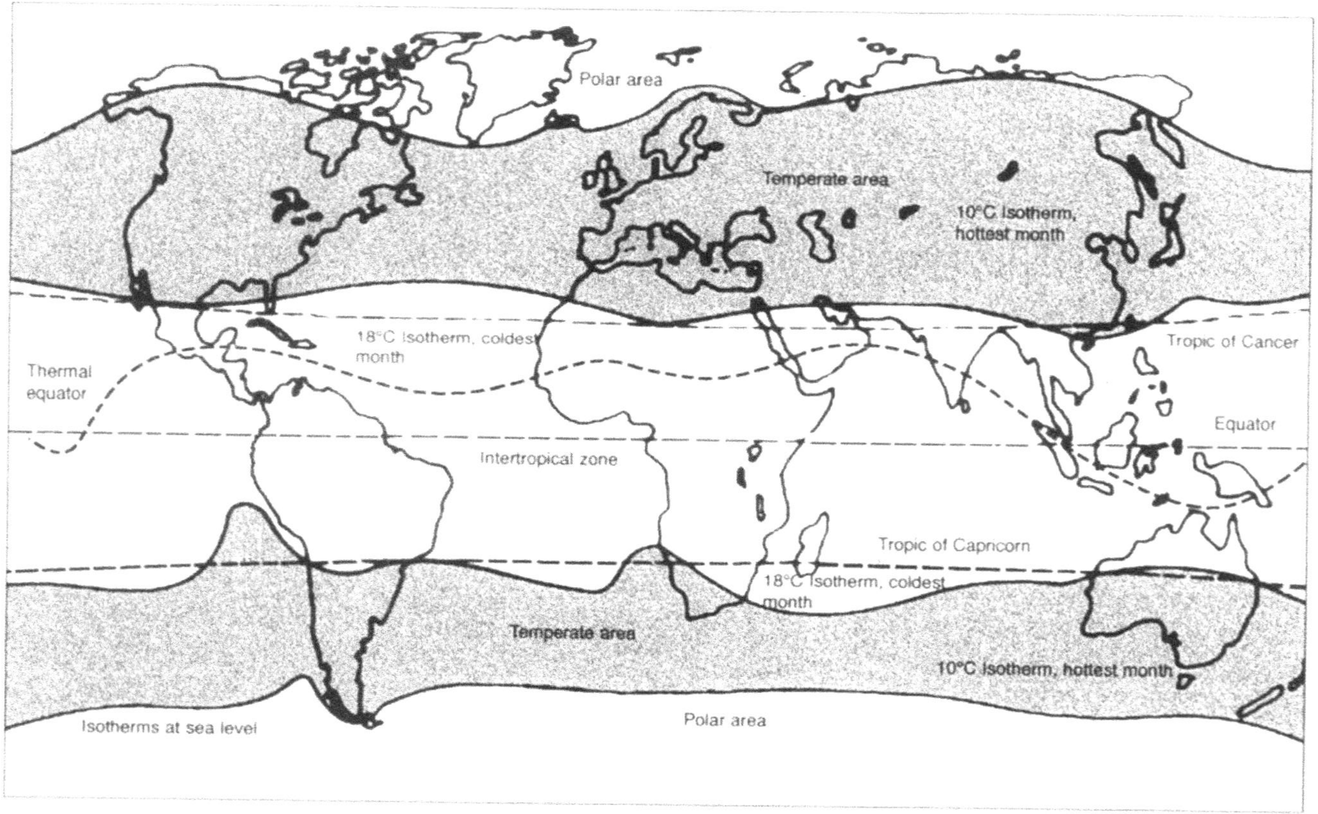

Figure 1 Climatic zones

Labels within figure:
Tropic of Cancer
Equator
10°C isotherm, hottest month
Temperate area
Polar area
18°C isotherm, coldest month
Intertropical zone
Tropic of Capricorn
18°C isotherm, coldest month
Temperate area
Polar area
10°C isotherm, hottest month
Isotherms at sea level
Thermal equator

6

o *The polar zones*
These are the zones outside the isotherms of the temperate zones.

This division is based on sea level (altitude 0 metres). At this level, the intertropical zone, which includes the majority of developing countries, is a warm zone.

However, relief plays a decisive part in determining temperature. In fact there is *a temperature fall of 6°C for every 1000 metres of altitude.*

Thus, at the outer limit of the intertropical zone, from 1000 metres altitude low temperature becomes a significant feature and the heating period is considerably extended. A climate is therefore defined as cold, when the average annual temperature is lower than 10°C.

The severity of a cold climate can be measured in relation to several variables:

o extreme temperatures, which can reach –50°C in the Himalayan regions;
o duration of the frost period – a characteristic of the utmost importance for agriculture;
o the very high daily variations in temperature in desert regions, due to the nature of the vegetation cover;
o significant annual temperature variations when the zone includes areas far from large thermally stable stretches of water; this is the influence of the continental land mass on the region.

From these variables it is possible to define several groups of cold climates in developing countries:

o *the polar-type climate*, which is found in very high mountains, principally in the Himalayan chain (India, China, Afghanistan, Nepal, Bhutan) and the Andean Cordillera (Ecuador, Peru, Colombia, Bolivia, Chile);
o *the subtropical dry winter climate* (also called sub-polar), which applies to the Chinese plateau and to the middle mountains of the Himalayan and Andean chains;
o *the subtropical dry summer climate*, characteristic of the Atlas mountains (Morocco and Algeria);
o *the dry desert climate*, which is found in the cold deserts (Gobi, Takla-Maklan deserts, part of the Iranian plateau and Patagonia);
o *the continental climate*, which includes the Caucasus and the Siberian plateau areas.

There are appreciable temperature differences within these large groups of cold climates. In addition, the concept of a microclimate is introduced in order to describe a local climate with sufficient precision.

This description only applies to some few square kilometres in mountainous areas, but such specificity is essential if the multiple characteristics of a site are to be understood. The microclimate can therefore be defined by the orientation of a valley, its vegetation, its position in the principal mountain chain etc. – so many factors which influence its amount of solar radiation, its exposure to the wind, the level of precipitation etc.

The climatic approach is thus essential in defining a cold zone. It allows one to know how to adapt the solar equipment (see Part Three) in the most appropriate way to a particular region. But another approach is needed in order to have a complete picture of a cold region in a developing country.

The socio-economic approach

Even if the cold regions are geographically very scattered, they do have a certain number of common features, which help to understand why they have remained marginalized. The principal ones are:

o *Minority status within the state as a whole, whatever the country concerned (India, China, South American countries)*
These regions contain high-altitude areas or desert space which are natural frontiers at the outer reaches of the central state. For this reason they are far removed from its economic and political influence and almost always far removed from its cultural centre (linguistic, religious, racial).

o *Limited agricultural resources*
Cold regions have to endure long periods of frost, which can exceed two hundred days a year. It is therefore impossible to work the land during this period. Furthermore, the arid nature of most of these climates limits agricultural development, which has to contend with poor soils and low rainfall.

Two types of agriculture are found in these areas: intensive, irrigated agriculture on small plots, and extensive, non-irrigated agriculture with very low levels of production (mixed farming, livestock).

o *Socio-economic systems based on subsistence*
The factors listed above have meant that it is essential for people to organize themselves to be self-sufficient. In economic terms they have developed a subsistence agriculture and internal commercial networks. In social terms, marriages are local and the birth rate is low.

o *Breakdown of existing structure*
In almost all cases the balance between environmental resources and the needs of the local population has been destroyed.

These regions have in fact experienced massive shifts in population or a significant increase in the number of inhabitants. Furthermore, they are experiencing a slow change in their ways of life as contact is made with the outside world, hastened particularly by modern communications.

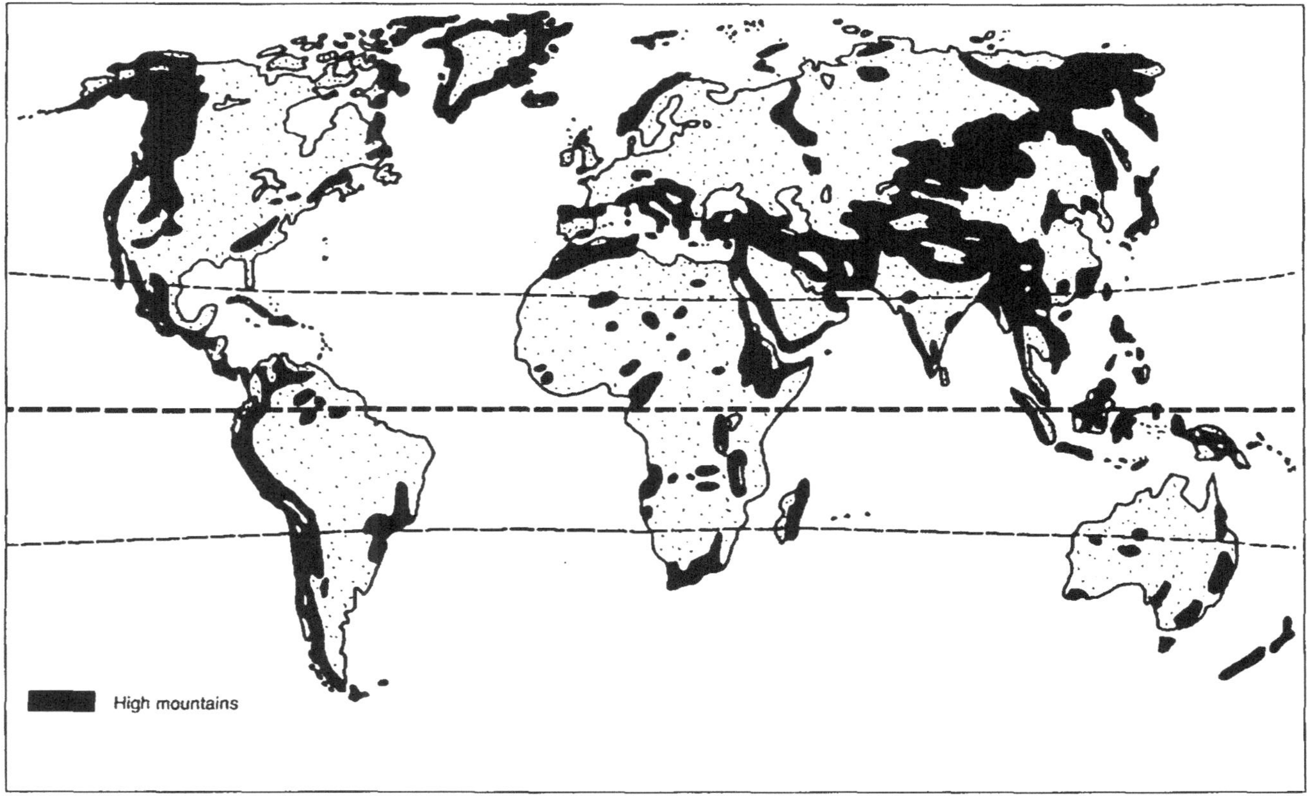

Figure 2 Areas of high altitude

High mountains

9

Eating habits have changed, new needs have appeared, economic exchanges have intensified and gaps in social status have opened up.

These remarks should not be taken to imply support for an idealized but archaic vision of the way of life. But they do underline that people in cold regions are confronted by increasingly serious problems in areas which concern them directly: health, education, nutrition, finance; or indirectly: deforestation and air pollution.

With regard to energy, provision for heating is one of the first things to fail.

The socio-economic effects of lack of heating

In some cases – schools and clinics in particular – the cost of fuel to maintain an acceptable temperature is beyond the financial means of local authorities. In some cases this gives rise to extreme situations. In rural areas, school children have to bring their own fuel to school; if they do not do so then the schools can refuse to allow them into classes.

In these regions the level of domestic thermal comfort bears no relation to that in cold or temperate regions in industrialized countries. During the winter season the indoor temperature is often below freezing. Heating is only available at meal times and energy used for cooking is recovered as far as possible. People are used to transferring the embers into the stove after cooking, or mixed systems are used for simultaneous space heating and cooking. Nevertheless, there is one exception to this very sparing usage: the indoor temperature level is maintained when a member of the family is ill.

Traditionally, fuel came from the immediate environment: bushes, brushwood, vegetable wastes, dried animal dung etc. However, with the increase in population, this is becoming unsustainable and is upsetting the local ecosystem. The result is increasing desertification.

XINJIANG, AN EXAMPLE OF DESERTIFICATION

Xinjiang, a province of north-west China, is an immense area of around 1 600 000 square kilometres. It has an extremely arid, cold climate with more than 200 days of frost per year.

The land is divided as follows:

Forests	17 300km²	1%
Desert	1 023 000km²	62%
Cultivated	58 700km²	4%
Mountains	545 160km²	33%

For 1984, the Chinese statistics estimated the supply of energy needs in rural areas to be the following:

Firewood	1 916 175 tonnes
Vegetable residues (straw)	2 971 400 tonnes
Animal (cattle) dung	1 578 000 tonnes

Because of excessive removal of material from the surrounding ecosystem, serious effects are being observed:

- o increased deforestation (removal of material is 1.7 times greater than the replanting rate);
- o accelerated desertification, sandstorms becoming more and more frequent;
- o increased time for gathering non-commercial fuel; the involvement of children of school age in fuel gathering; marked deterioration in the conditions of life in rural areas. Having an annual income of 660 French francs ($132 – average estimate made in 1991), the peasant in Xinjiang is not able to obtain other sources of energy. The increasing cost of the coal produced in the region causes severe heating problems.

On the other hand, the use of fossil fuel in areas of high population density has given rise to an alarming level of atmospheric pollution. In China, where people burn half a kilo of coal per head daily for heating, because of the 200 days of frost per year in the most severe cold climates, the use of locally mined coal (95 per cent of commercial fuel used) is the main cause of this type of pollution.

Energy issues are also economic in nature. Artisan and agricultural activities are reduced during the winter because the artisans and peasants cannot meet the high costs of commercial fossil fuel (oil, kerosene, gas, fuel oil, coal etc.). Agriculture can even be completely paralysed because of the associated transport costs. For the same reason, imported food remains beyond the reach of rural people.

The situation in Ladakh is indicative of this state of affairs. A high valley at over 3000 metres altitude in the Himalayan chain, it is surrounded by passes of up to 5000 metres, which are the only access for road transport. During the winter these access roads are blocked by snowfalls and the only means of communication is by air.

The cost of fresh food brought by plane is beyond the means of a Ladakhi peasant. His only chance of survival during this season is to make use of food stocks accumulated during the summer, and to hope that the winter will not be too long. He therefore not only has to put up with the severity of the climate, but also with the problems of food shortages. A Ladakhi proverb neatly sums up this situation of isolation: 'An adult Ladakhi was born in summer.'

In the agricultural sector attempts have been made to try to overcome this winter paralysis but, as they require commercial energy, they have usually failed.

For example, some have thought of setting up kerosene incubators to make the hatching and rearing of chickens viable in the severe climate of the high valleys of Peru. But the incubators have fairly rapidly run into two types of problems: poor management of the technology, resulting in frequent fires, and excessive operating costs. It could be said that the large budget needed for fuel has impeded development of this technology.

Tunnel greenhouses inspired by systems used in industrial agriculture also failed. In fact they required substantial complementary heating, which,

11

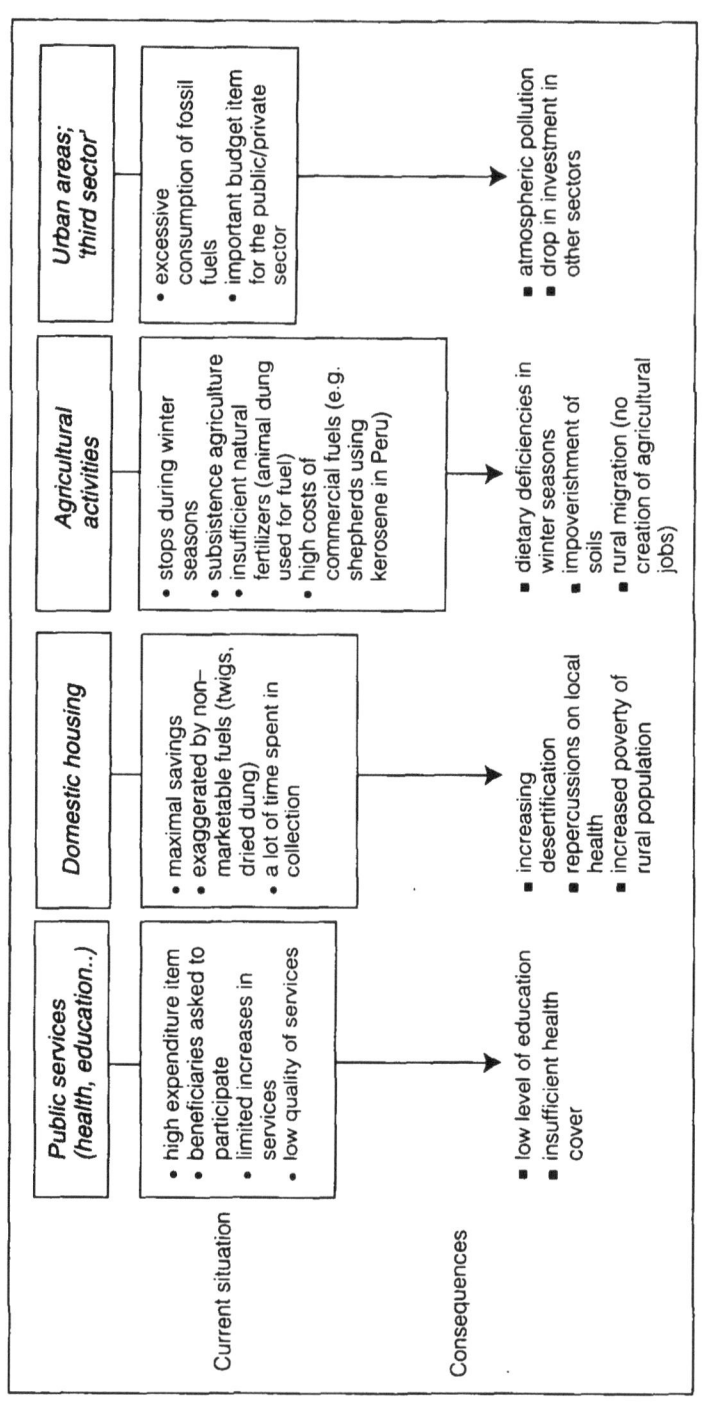

Figure 3 Energy needs for heating

12

as in the case of the kerosene incubators, was inappropriate in the economic context of cold regions of developing countries.

This overview of the issues relating to energy in cold regions shows that setting up heating systems in these regions must meet the following criteria:

o low purchase, operation and maintenance costs
o easily managed technology
o adaptability to the needs of the people concerned
o local availability of materials for construction
o absence of air pollution problems.

Why passive heating is suitable for the cold regions of developing countries

Active and passive methods

The principle of solar heating is simple. It consists of taking advantage of solar radiation and converting it into heat. To achieve this, all that is needed is to place some dense material in the path of the radiation so as to convert it into heat. Once that has been achieved, there are two possible ways of using the resulting heat:

o The material is moved to the space to be heated where it releases its heat. This is the active method.
o The material is already in contact with the space to be heated and releases its heat to it. This is the passive method.

In the case of the active method, water or air panels are generally used, which absorb the heat and then transport it via a conduit to the space to be heated. In the case of air panels a fan blows the air directly into the space or puts it indirectly in contact with the space via a dividing wall, for example.

For water-filled panels, a pump drives the water in indirect contact with the space, via sloping convectors in a partition wall or in the floor for example, or it feeds it directly, as in the case of a swimming pool.

It should be noted that active systems require a mechanical means of transferring the heat.

The passive method works in a completely different way. It demands careful design of the space to be heated. It consists of subtly using the choice of construction materials and making appropriate shapes according to the outside climate. An internal microclimate is thus created. This is the case with igloos, caves dug from the rocks or underground (troglodyte shelters).

The 'greenhouse effect'

The modern design of bioclimatic housing relies on the use of transparent materials, principally glass, which allow solar radiation to pass into the space to be heated and then trap the energy.

It should be remembered that when solar radiation enters a space through a glass panel, a large part of its energy is converted to thermal radiation on contact with dense matter.

Now, transparent materials such as glass are completely opaque to this type of radiation, which is trapped in the space, thus retaining its heat: this is the greenhouse effect.

The space carries out the functions of collector and storage of solar energy on its own, in order to cover the heating requirements over 24 hours.

This method has multiple applications for housing and agriculture.

In the case of housing, heat can be collected and retained in the house thanks to:

o a solar wall;
o a greenhouse veranda;
o systems combining different heating methods.

In the case of agriculture, it is possible to use:

o various solar greenhouses;
o special systems for hen houses;
o systems designed for various purposes, for example for mushroom cultivation or for composting.

We will develop all these examples in Part Two.

The advantages of passive solar systems

Passive solar systems combine a number of qualities which make them easily adaptable for people in cold regions:

o *They use energy which is free and available everywhere*
The majority of cold regions benefit from a high and constant level of insolation, which is an essential condition for good performance of passive solar systems. This is due to the severe climates with dry winters and rare cloud cover.
Solar energy is the only source to have these two fundamental qualities, whereas other energy sources have high purchase and operation costs.

o *Passive solar systems require a lower investment*
Because they do not rely on mechanical systems, people do not need to import expensive precision components, or call upon technicians from outside.

14

o *Passive solar systems can be self-built*
Because of the simple techniques required and the low investment necessary, users can build them for themselves, and construct them on site.

o *Passive solar systems require little maintenance or monitoring*
The basic principle is that they work unattended. They only require simple operation, with opening and closing of shutters, windows and doors, morning and evening.

o *Passive solar systems can be adapted to existing buildings*
This advantage is fundamental in developing countries where householders are often short of resources. They do not have to invest in the construction of the living space to be heated. The systems merely have to be positioned on sunny parts of existing buildings.

On the other hand, the masonry parts of vernacular buildings in cold regions are characterized by high thermal inertia. This can be used to advantage to store heat.

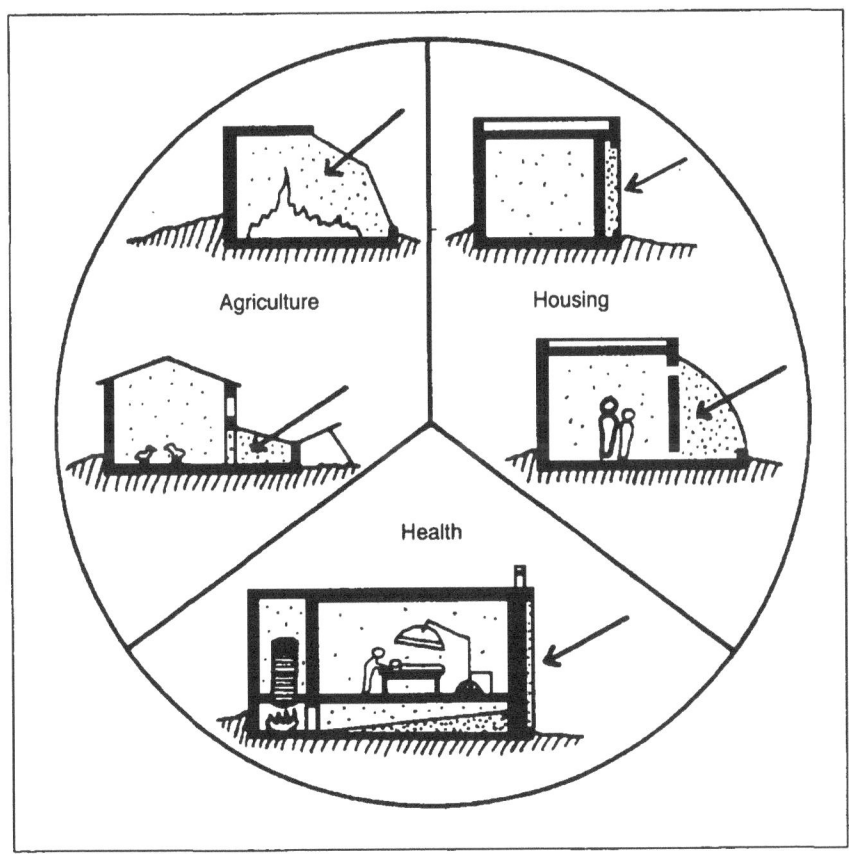

Figure 4 The various applications of solar heating

15

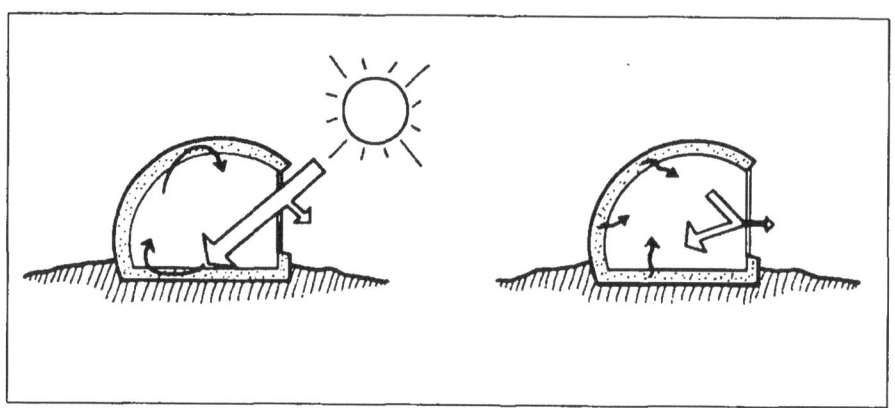

Figure 5 The greenhouse effect

o *Passive solar systems can be part of a mixed heating system and meet existing criteria for comfort and hygiene*
In this document we present the example of a workroom whose temperature level and quality of interior environment are provided to perfection. There is no pollution from smoke emissions and dust infiltration is effectively controlled. This performance is possible by combining a solar wall and heating the ground by the *hammam* principle, that is to say using the heat transmitted by the exhaust gases from an external hearth.

o *Because of their reduced costs, passive solar systems help create new agricultural activities*
The examples of greenhouses and solar poultry sheds presented in this book are revealing. Employment and income have been generated by using them while technologies based on the use of fossil fuel have been displaced. These systems have also helped to create new jobs such as market gardening under glass, which was previously unknown in a rural area. They only require a small area and maintenance work which can be done outside normal agricultural working hours.

o *Passive solar systems are a response to the problems of atmospheric pollution and reduce the greenhouse effect*
The use of a greenhouse-effect wall during the heating season produces 150–200 kWh/m^2 (kilowatt-hours per m^2), which is equivalent to 200kg of CO_2 per year released into the atmosphere to produce the same quantity of energy from coal.

16

Experimentation and applications

The four cases described below are representative of cold regions in developing countries. The research in each case has provided a good illustration of the challenges offered by solar energy. We briefly outline the most striking features of these experimental situations.

The Moroccan Atlas

The research carried out by Faïjal Benchekroun[*] made it possible to analyse the consumption of wood fuel in rural areas of the Middle Atlas.

It emphasizes that geographical situation has even more significant impact than socio-economic characteristics on heating needs. The Moroccan Atlas has therefore been divided into four strata which take into account the various altitude levels of the region which determine the periods of cold and thus the heating period, that is to say the firewood consumption. Thus:

Zoning of the Atlas (according to Faïjal Benchekroun)

Stratum 1: altitude over 2000 metres

The heating period lasts on average 130 days per year. The consumption per inhabitant is of the order of 3.1 steres per inhabitant per year (stere of firewood = 1 cubic metre of wood).

Stratum 2: altitude between 1000 and 2000 metres

The heating period is 95 days per year. Consumption is of the order of 2.8 steres per inhabitant per year.

Stratum 3: altitude between 500 and 1000 metres

The heating period is around 90 days per year and the consumption 2.4 steres per inhabitant per year.

Stratum 4: Altitude below 500 metres

The heating period is no more than 75 days per year and the consumption is 1.9 steres per inhabitant per year.

When the household is less than five kilometres from a forest, the head of the family or one of the members of the family is charged with going there to collect firewood.

When this distance is over five kilometres the problems of supply are more acute. In disadvantaged sections of the population one member of the family must undertake this task on a full-time basis, and this disrupts

---

* Hassan II Agricultural and Veterinary Institute, BP 6202, Rabat Institutes, Morocco

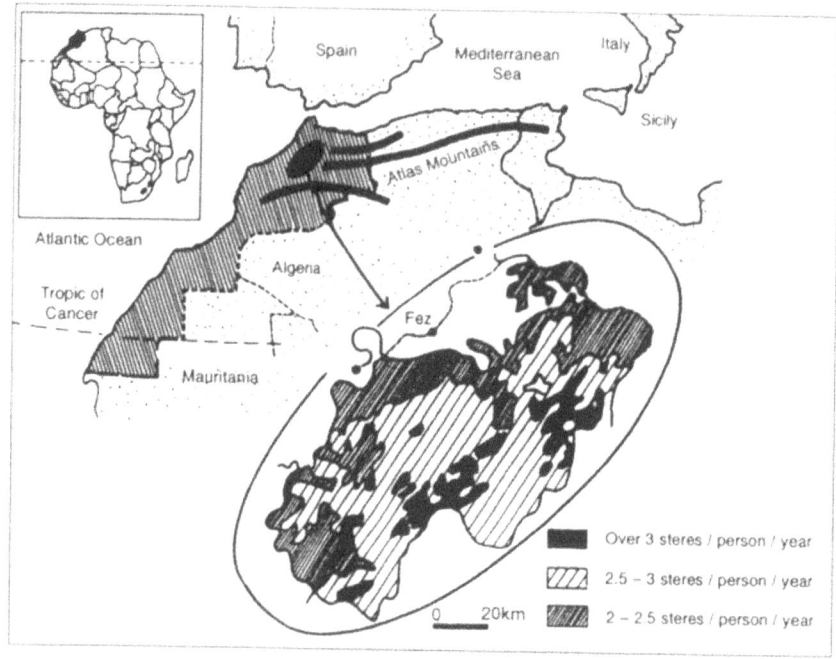

Figure 6 Firewood consumption in Morocco

farming work. And even then the supply of wood is not assured on a regular basis. Families often have to be satisfied with eating cold food, and malnutrition results. In fact, only rural people with a high income do not suffer from lack of wood during the year because they can always cover the cost (around FF1/kilo = US15c/kilo).

When the distance from the fuelwood source is over 12 kilometres, as in the case of villages in the cold steppes, a regular fuelwood supply is almost impossible. Even the most disadvantaged families must find significant sums of money to procure wood as well as making maximum use of substitute fuels such as vegetable waste and cow dung.

Nevertheless, wood remains the only reasonably accessible means of heating for disadvantaged rural people. This has certain consequences for the environment, to such an extent that some have said that the primary cause of desertification in Morocco is poverty. In this country the deficit in firewood has been estimated at between 3 and 10 million cubic metres per year.

This is confirmed by the work of Laurent Auclair in the High Atlas (1991): 'During the cold winter days the consumption of firewood is approximately double [that of the period when heating is not used]: 18 to 35 kilos per household per day according to the families. On average, according to the mountain villagers, heating requires between 1.2 and 2.2

tonnes of wood fuel per household per year, assuming 100 cold days per year.'

To summarize, the research of F. Benchekroun shows the many factors at stake in the use of solar energy: the struggle against desertification in mountainous areas, improvement of living conditions in terms of thermal comfort, savings in firewood collection times etc.

Finally it should be emphasized that the use of solar energy in Morocco is particularly practicable because the traditional construction, with thick *pisé* walls (rammed earth made on site) and heavy roofs, provides a high level of [thermal] inertia. Passive solar heating systems are therefore particularly appropriate to the Middle Atlas of Morocco.

Sikkim (India)

The research by P.P.S. Gusain (the Development Alternatives Association)[*] provides a basis for energy planning, particularly for renewable energies: solar, biomass, water and wind power.

The results of this research have similarities to the Moroccan case. On the one hand the consumption of biomass greatly increases during the winter; on the other hand this consumption depends on the economic level of the family, with more biomass being used by poorer families.

In Sikkim, the work of firewood collection also makes up a significant part of the annual labour of an adult rural dweller. It has been calculated that this activity takes up 90–150 days per year, two to six hours per day.

This study also made a link between the smoke emissions from biomass combustion for heating and cooking and the state of health of users. This correlation is confirmed by the work of LNP[†] in Ladakh, which noted an improvement in health when improved stoves were disseminated. The graph on page 21 illustrates the results.

As in the case of the Moroccan Atlas, the main issue in using solar energy is the struggle against deforestation. Over six years the time spent collecting firewood has increased from an average of two to three hours per day to five to six hours for families with five to six members. In winter, the use of agricultural residues does provide for energy needs, but it results in a corresponding reduction in their potential use for enriching the soil, and thus reduces productivity.

It is estimated that, with an average annual population growth rate of two per cent, in spite of improved living standards and greater utilization of commercial fuels (kerosene, electricity), the need for firewood supply will become increasingly acute. The shortfall in supply of firewood will

* Development Alternatives, an Indian association specializing in technology transfer, B-32 Tara Crescent, New Mehrauli Road, New Delhi 110 016, India.
† Leh Nutrition Project, a Ladakhi association working for health improvement, Kham Manzil Zangsti, Leh, 194 101 Ladakh, Jammu and Kashmir.

Figure 7 Location of Sikkim

continue to grow in spite of the substantial increase in tree plantations and the use of improved stoves.

The report recommends that to meet increased needs for firewood all possible means of responding to energy demand should be mobilized. This means an increase in supply from tree planting, specially selected for this purpose (rapid growing, low toxicity smoke, high calorific value, rustic nature) together with fuel-saving measures at all levels: use of improved efficiency stoves and cookers, good plantation management, and above all, use of solar energy. In Sikkim, as in Morocco, the use of passive solar heating systems would reduce firewood demand.

20

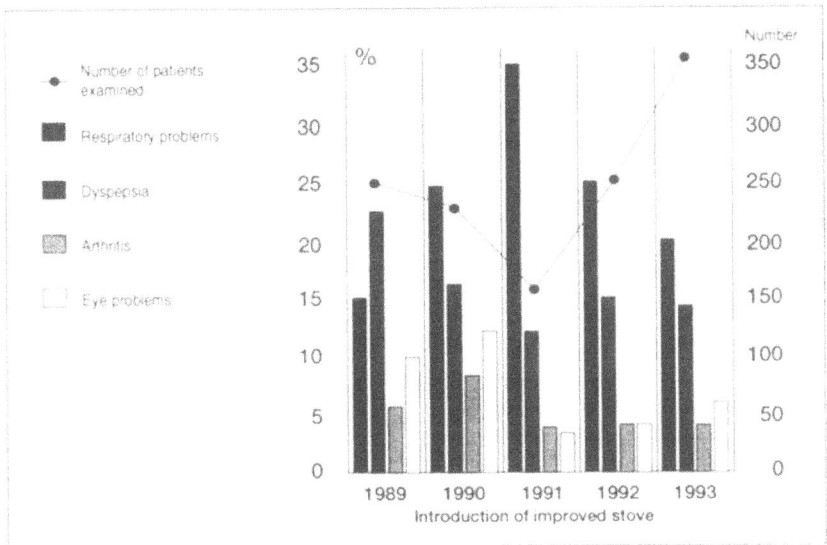

Figure 8 Health impact of smoke reduction
The distribution of disease following introduction of improved stoves from a report by the Leh Nutrition Project, January, 1993

Peru

The Runamaqui Association* at Anccopaccha in Peru has decided to use solar energy for agricultural purposes in the region. Besides its usefulness for household heating, the use of solar energy for agriculture is opportune, particularly if the technical choice improves its economic viability.

The following passage is an extract from the book *Soleil du Sud* published by the association following this research project (see also Bibliography).

CHICKEN REARING IN THE ZONE OF ANDAHUAYLAS

An explanation of the issues requires a brief overview of chicken rearing in the Anda-huaylas region in 1974/75.

From its domestic origins, poultry production in the Sierra started with chicks which were already hatched. Neither incubation nor hatching took place in the mountains.

The conditions at about 3000 metres are unsuitable for reproduction of either local breeds or industrial hybrids. The oxygen pressure is too low for it to diffuse through the eggshell and incubation, even on a reduced scale, would present prohibitive technical problems for these regions.

Day-old chicks are therefore ordered from the incubators on the coast and

---

* Runamaqui, a Franco-Peruvian association, rural development projects in Peru, 8 rue Bezout, 75014 Paris, France

21

Figure 9 Location of Peru

transported by air and/or truck. These are selected hybrids (meat chickens, egg layers or a mixture) and the necessary feed and vaccines for rearing are also purchased at the coast. Because of low outside temperatures, the chicks have to spend their first three or four weeks being reared. This is the initial rearing. After this first phase the chicks are stronger and can be kept in the farmyard. They are then sold at the local market at three to four weeks.

The buyers are for the most part local people who want to raise the chicks for eventual consumption, but also peasants, almost all of whom practise *la recria*, domestic rearing in the farmyard for several months to re-sell the majority of the adult birds in turn at the market.

The real return on this *recria* is often very low, but it does enable local people to obtain money for a very low investment. For peasants it provides a secondary source of income. For this reason, the high mortality of birds during *la recria*, due to the often very inadequate initial rearing stage, does not call the practice into question.

22

Practised by the urban middle classes, the initial rearing is for them a secondary activity which is terminated as soon as it becomes less profitable (costs fluctuate considerably), or supplies (chicks, feed etc.) become difficult. In good times, the Andahuaylas region has a maximum of three or four rearers, each rearing between 1000 and 3000 chicks. No rearer active between 1976 and 1980 was still active between 1984 and 1987 except for Mario, a technician at Runamaqui.

The initial rearing carries certain risks which are a consequence of the problems of energy supply.

In some cases the local heating of the rearing room is provided by a ring of light bulbs, held at floor level; this system does not permit effective temperature regulation, particularly because of the frequent power cuts during the night.

The current technique is the diesel fuel incubator, consisting of a burner, covered with a sort of lampshade, which gives it the appearance (and the name) of a bell (*campaña*). Sited at the centre of the room some forty centimetres from the floor, it creates a warm area where the chicks congregate if it is too cold in the rest of the room.

Obviously it is in the interest of the rearer to use as little fuel as possible, and therefore to keep the temperature at the minimum. The chicks are thus almost always under the *campaña*, sending up clouds of dust from the wood shavings that cover the floor when they move about. *On average one brooder per year is destroyed by fire (out of seven or eight)*. One rearer has solved the heating problem in an innovative way: he raises his chicks in the Rio Pampas valley at a low altitude (around 1500 metres) where the temperature is tropical. But he has journeys of five to six hours.

The idea of designing a *campaña solar* (a solar brooder) thus offers several advantages:

○ energy self-sufficiency means that an activity formerly restricted to towns in the valley may be established in rural areas;
○ there is no fire risk;
○ if the installation is properly sized, it is possible to achieve a higher average room temperature than that achieved by conventional systems, with no additional running cost. Thus feed consumption can be reduced.

Chick rearing in the Sierra communities is dependent on three vital areas of supply. The first of these is the supply of *energy*; the others are the supply of day-old *chicks* from Lima and the supply of *feed* for the chicks.

In this context, the use of solar energy remains the best alternative in the face of problems encountered in chick rearing. Without ignoring other aspects mentioned (feed and transport of young chicks), the use of solar energy for heating rooms proves to be the essential ingredient for successful chick rearing in high-altitude rural areas.

Ladakh (India)

The following is an account of the experience of GERES in Ladakh. It concerns the distribution of solar greenhouses in the high valleys of this Indian province.

Initially the main aim of installing solar greenhouses was to limit food shortages during the winter season by encouraging the growing of leaf vegetables such as spinach. It was also to allow an increase in production

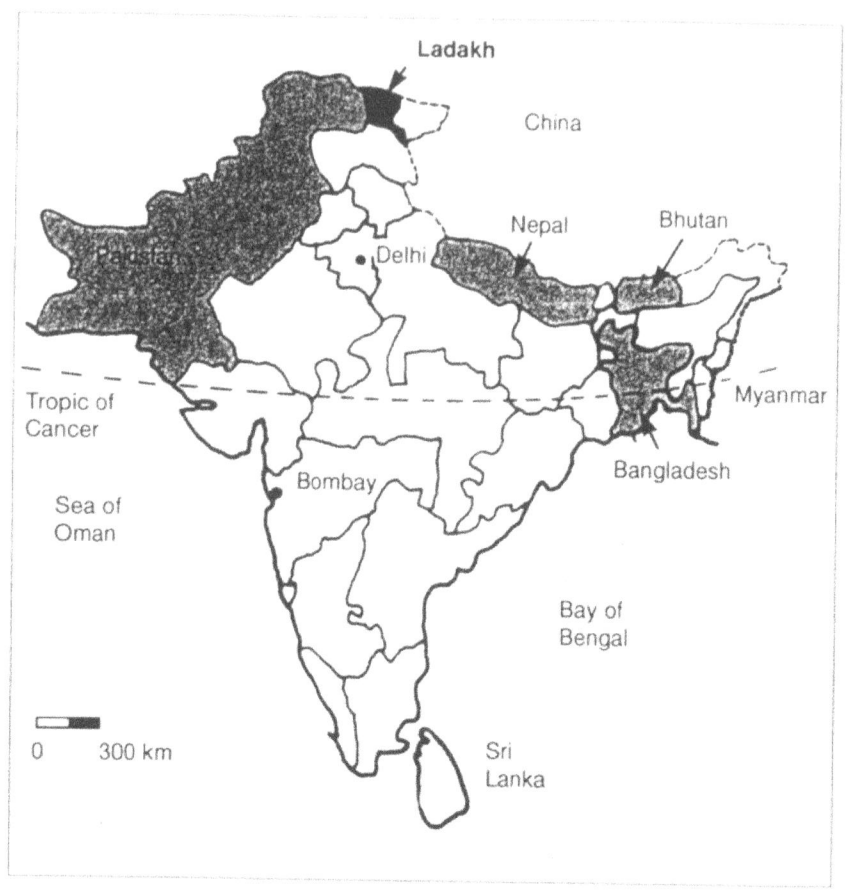

Figure 10 Location of Ladakh

through planting out under glass, as soon as the soil was frost-free, thus permitting two harvests per year.

But this solar equipment can also be used for a commercial purpose. The greenhouse produce is sold locally if outlets exist, and thus becomes a significant source of revenue for the population.

Through the example of the village of Thiksay, situated 20km from the capital, Leh, economic factors in the use of solar greenhouses can be measured.

Because of its position on the banks of the Indus, Thiksay enjoys large agricultural areas and abundant water.

The landholdings average 1.3 hectares and each family (seven people on average) has to secure food requirements from these sources. However, in most circumstances, the area used for market gardening is very limited. Use of greenhouses does allow an optimal usage of the growing area, to a maximum of 10 per cent of the total cultivated area.

24

It has been estimated that one third of families possessing a vegetable garden (here meaning a vegetable plot not used for growing potatoes) would like a greenhouse. Before acquiring one, potential owners have to calculate what investment they can afford and their level of technical knowledge of vegetable cultivation under glass.

It is also necessary to know whether this system is financially viable. In 1993 it was found that efficient use of nine square metres of solar panels produced an annual income of 17 000 rupees (US$1 = 25 rupees). If a worker's pay is of the order of 10 000 rupees, the use of panels enables one person per household to live full time from its use. In other words this can mean that 20 per cent of employed people now working outside the village could find a job at home in market gardening.

Using solar panels in rural areas becomes a way of effectively combating unemployment and the rural exodus, by stimulating economic activity. What is more, as they only cover a limited area, they are not in competition with traditional production.

Moreover, there are obvious outlets for agricultural production under glass in the region. The Indian army with its high profile has in fact huge requirements, of which Ladakhi production can meet only 5 per cent.

It can be seen that solar greenhouses can introduce new agricultural methods into Ladakh. Their scale of operation is adapted to an economic development which does not displace traditional farming practices. Farmers have very readily accepted it, as demonstrated by a rapid growth in demand. At present there are more than 500 solar greenhouses per 150 000 people in Ladakh.

Conclusion

The different case studies show that using passive solar techniques seems to be a way of solving energy problems particular to cold areas. They are not presented as a miraculous solution but to show through the experience gained that they are one of the essential ingredients for improving the living conditions of people living in these areas.

2 An intervention methodology

This book should enable the reader to grasp the possibilities for using solar energy in cold areas, to understand the main physical phenomena concerned and to develop strategies for action which take account of local conditions. The reader, whether a development worker, programme decision-maker or whatever their role, must always take into account the constraints associated with any technology transfer. They should be particularly careful about cultural differences and the living conditions of people in these regions.

Among the special characteristics of these regions are their total geographic isolation for several months of the year, which gives rise to specific cultural differences. What is more, these characteristics are often positive integrating factors for development activities.

Thus it is noticeable that isolated rural people (in mountain valleys) often prove most capable of setting up organization systems for the management of scarce resources such as wood and water. This is true in many parts of the world but in order to establish activities introducing solar devices, it is necessary to have a clear understanding of the mechanisms which govern organizational methods within each region.

Equally it is important to bear in mind that the majority of traditional rural organizations are disintegrating as a result of the rural exodus, pressure on the land because of overpopulation, the influence of urban lifestyles etc.

Some of the techniques presented in this book have been designed in an urban environment (such as hospitals, schools, private houses etc.) where ways of disseminating knowledge are very different from those found in rural areas. To enable these technologies to be adapted a series of questions has been prepared, the answers to which will help to provide an effective procedure for the dissemination of solar techniques.

EXAMPLES OF QUESTIONS TO BE ASKED ABOUT AN ECONOMIC ACTIVITY (e.g. market gardening under glass, poultry rearing)

- Are there local examples of similar ventures?
- What are the traditional knowledge/skills in the area?
- What training needs (technical, management, literacy etc.) would there be?
- What existing training networks are there?
- What social organization, social structure, investment capacity is there?
- Is there an existing traditional or institutionalized banking system to provide credit?

○ Is family, group or individual activity feasible?
○ What are the features of the public and private sectors, and how do they relate?
○ What is the most appropriate part of the community with which to work?
○ Can social conflicts be anticipated?
○ How are marketing networks (market, business centre), preservation, storage, transport and distribution methods organized?
○ Who is responsible for sales and how are prices determined?
○ What are the social motivating factors (increased income, social recognition, training)?

This is not an exhaustive list, but it does already help to outline the complexity of an existing situation.

It would be misleading to suppose that the introduction of new technology will not disturb the recipient socio-economic system. It is therefore essential, having carried out in-depth analysis of local conditions, to define the difficulties with regard to suitability (including cost of systems, training needs, responses to needs, possibilities for replication etc.) and to anticipate future changes which are bound to occur: the development of new commercial networks, changes in social structure, lifestyles etc.

Here an overall framework methodology for intervention is being presented. It is part of a technology transfer model which is applicable to every type of technology innovation programme.

Methodology for a programme of dissemination of solar techniques

It is important to remember that the objective is not simply to disseminate solar techniques but to transfer the associated technology (design, construction, use and adaptation). This means that there must always be a connection between the dissemination of these solar techniques and the needs of the beneficiaries, their capacity to invest, their existing knowledge and skills etc. A clear grasp of these points is essential for the success of the programme.

The methodology has three stages:

1. Gathering data and identifying obstacles for a programme of action.
2. Implementing and monitoring development of the programme.
3. Evaluation of the research project so that it can be repeated in the future.

Stage 1: Data analysis and programme design

There are various ways of gathering data. Several schools have developed their own methods of obtaining the most realistic picture possible of the environment being studied. These methods are often derived from social science methods (sociology, anthropology etc.).

It is important to remember that information will be more accurate if it originates from the organizations and structures which are an intrinsic part of the local environment.

The following provide an overview of basic information to be gathered so as to define the essential characteristics of the environment, to highlight the intervention priorities and to identify potential partners for a future programme.

Information gathering

There are many types of information to be collected, in fields as diverse as technology, sociology and economics.

o *Population*: demographic data; racial, social, professional and religious composition; traditions, lifestyles etc.
o *Health*: number of households, urban/rural split; physical facilities, energy problems (including heating).
o *Education*: school attendance levels, cost of schooling; problems of space heating; other problems.
o *Agriculture*: type of agriculture (subsistence, market); size and common characteristics of farming practices; crops grown, marketing networks.
o *Public authorities*: government support; national or regional policies for renewable energy for agriculture, domestic use, health etc.; priority issues for action; relationships between public and private sector.
o *Financial systems*: national and regional banking arrangements; local organizations (credit and savings unions); traditional systems; inflation; favoured sectors, etc.
o *Management positions/skills networks*: identifying people and networks which are qualified and motivated to support the proposed programme; reviewing local experience in solar energy; possible training units.
o *Technical information*: knowledge of the microclimate, meteorological data from nearest meteorological station, wind regime and insolation of possible sites, current state of development; construction techniques, locally available materials, possibilities for supply of other materials (glass, timber etc.).
o *Overcoming constraints*: an intervention cannot always be justified on its own merits. There can also be prohibitive obstacles to the diffusion of solar energy systems.
o *Technical impossibility*: excessive cloud cover (lasting more than four days) during the heating season; solar radiation blocked by natural obstacles (mountains etc.); insufficient period of use to justify the initial investment.
o *Contextual difficulties*: all factors outside the control of local operators:

crisis situation, emergency, national programme which does not favour renewable energy use etc.

Planning activities
Having confirmed the starting conditions for the project, it only remains to define the framework for activities.

Firstly, systems have to be chosen. The choice takes account of the technical and socio-economic information gathered.

Next, targets for dissemination of systems should be set: in other words one must define at whom they are targeted; one must know how they can be installed, with what level of participation (financial, hardware or human) on the part of the beneficiaries; one must also know in what training networks, and according to what work practices, they will be disseminated.

The important thing is not to disseminate large numbers of systems, but to provide an enabling environment in which dissemination will occur, by providing all the means of acquiring the technology, controlling costs and checking people's interest.

Stage 2: Implementation of the programme

There are two dissemination phases.

The first phase acts as a test phase to evaluate whether the technological choices made are well suited to the capacity of the local operators.

During this stage, small-scale dissemination is carried out and operators use the new techniques for a year. This allows time to make any necessary modifications. It is recommended that several variations of the same model should be tried so that a choice can subsequently be made, after they have been observed in practice.

This first dissemination stage is essential in order to field-test the different components of a programme. It enables technical solutions to be speedily readjusted. It allows follow-up management and training structures to be finely tuned and it permits assessment of the impact of the programme on the people involved and of their acceptance of it.

After the limited dissemination phase one can go on to larger-scale dissemination according to the objectives laid out in the schedule. This is the second stage of the programme.

Stage 3: Evaluation and future

This is the final stage in the methodology. It will enable activities undertaken to be evaluated and strong and weak points to be highlighted, all with a view to obtaining maximum benefit from the experience. This critical analysis assumes that evaluation criteria have been defined at the outset.

It is very often found that those in charge of programmes do not attach sufficient importance to evaluation, although it provides a considerable amount of information and helps to measure the impact of a knowledge transfer project in concrete terms.

The progress of the Runamaqui association in Peru with a solar chicken rearing project provides a good example. Having carried out their programme they wanted to measure the impact. For this purpose they drew up an evaluation matrix, which can be applied to all types of project and which takes into account six major parameters:

1. the success of the project according to those involved;
2. the effects on the attitude and thinking of those involved;
3. its contribution in terms of acquiring knowledge;
4. the technical viability of the project;
5. the economic viability of the project;
6. its dissemination.

PART TWO

NINE SOLAR INSTALLATIONS FOR HOUSING AND AGRICULTURE

Here we present nine reference models which have been installed in Ladakh (altitude 3500m, latitude 34°N, longitude 77.6°E) for a long enough period – ten years – to be sure of their reliability.

Standardized diagrams are shown with the specific characteristics of each model. Each can be used as a reference for other areas with comparable situations, but in no case is there a rigid design framework for similar applications. Thus it is up to the designer to adapt these models according to the required results, the available materials and the local construction techniques.

The examples are illustrated by the simulation curves (software: Comfie, cf. Part Three, Chapter 10) for which the calculations have been validated by measurements made in Ladakh in February 1993. This allows the effect of the various parameters on the thermal behaviour of the presented systems to be evaluated.

There are two applications of the systems presented – housing and agriculture.

The models shown below are divided into these two categories. It will be noted, however, that there are similarities between the models presented in the two categories. For example, the adjoining veranda type (Case 3) and the open field greenhouse type (Case 5). In this case, to avoid giving the same explanation twice, the characteristics of one can be used for the two models.

3 Housing types tested in cold regions

Case 1: The solar wall

Construction principles and use

It is well known that a sunny unglazed wall, painted a matt colour, will store a little heat during the day, which is rapidly released to the exterior.

The principle of the solar wall is to place a sheet of glass between the outside environment and the exterior wall of the building. Two effects combine to increase the efficiency of heat storage in the wall.

The collector effect
The wall, warmed by the solar radiation, emits thermal radiation which is trapped by the glazing, which is opaque to the further infra-red (thermal)

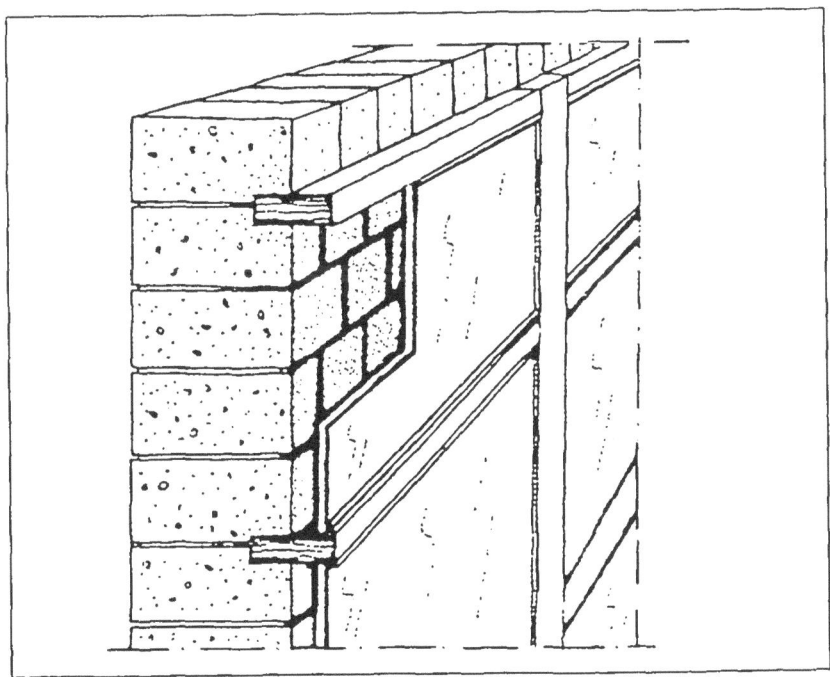

Figure 11 A glazed wall

radiation. This is the 'greenhouse effect' of the glass. The wall remains insulated from the climate outside (wind and temperature) by the glass covering.

The outside surface temperature of the wall thus reaches 60 to 80°C in the daytime.

The storage–release effect

The solar wall both collects and stores energy. It provides for the release of stored energy to the interior of the building after a period of time.

The time taken for heat to reach the inner surface is called the *lag*.

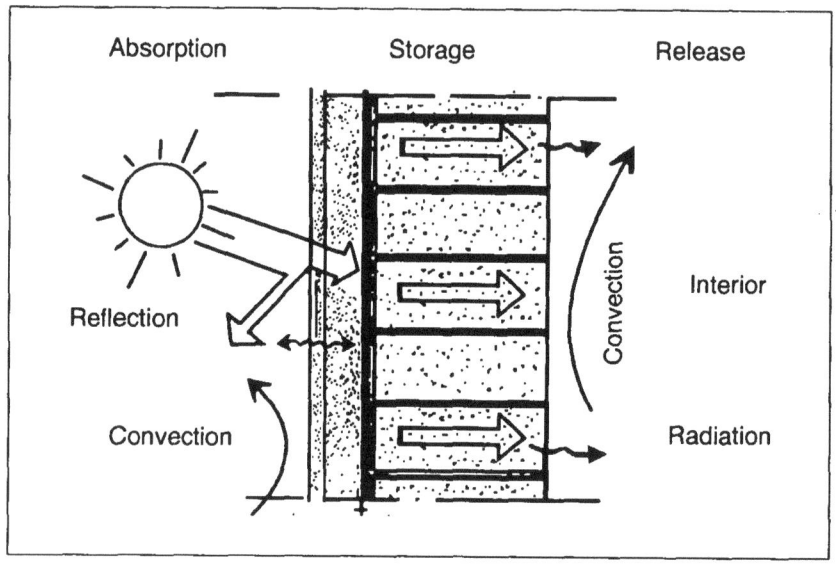

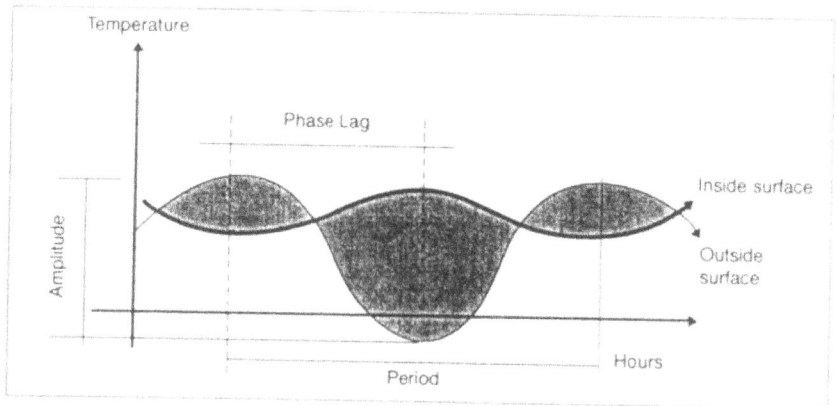

Figure 12 Delay between the absorption on the outside surface and release at the inside surface

34

Thus the solar wall is a system of delayed heating, allowing the energy needs of the main rooms of the house (living room, bedroom, etc.) to be maintained beyond the daylight hours.

The term 'collector–accumulator' is also used to denote the combined roles of this process.

From inside to outside the house the solar wall comprises:

o a wall made of 30cm × 5cm × 15cm mud bricks, placed perpendicular to the line of the wall and with each row offset, one above the other. The 30cm-thick wall is constructed in this way and painted matt black on its outer surface.

o double glazing with 1.25cm or 2.5cm spacing, depending on the dimensions of the dividing battens available. These battens must be carefully positioned so as to form a dust-proof seal over the whole surface of the building, while leaving sufficient play (2mm) all round the glass panes so that they do not break because of thermal expansion.

o a 5cm × 15cm section wooden frame, anchored in the wall at a depth of 7.5cm or greater if possible. This frame forms a trellis on which the glass is placed.

The following describes a house of modest dimensions (Mrs Spalzon's house), which uses a solar wall with three windows. It is in fact essential to have a glazed part so that there is still enough light inside, and direct energy gain can take place when the room is being used during the day.

Installation

The wall

This must be built of dense material (concrete, rammed earth, etc.) with good thermal conductivity. Earth is often a good compromise, in the form of *adobe* blocks, or earth rammed on site (*pisé*). It is often the preferred material for traditional housing in harsh climates.

The recommended thickness for earth is 25 to 35cm. If it is used in the form of adobe blocks, only one layer is used. This facilitates heat conduction to the interior of the wall. Having space between two materials adds an extra resistance to heat transfer.

Use of an existing wall is possible if its thickness and composition are as recommended above.

The horizontal and vertical jointing are done with as thin a layer as possible of a more conductive, fine-grained grouting compound.

All that remains then is to blacken the outside of the wall to increase its absorption. A traditional coating, with black soot (carbon) added, gives good results.

Careful monitoring of the quality of the outer coating is important as it is

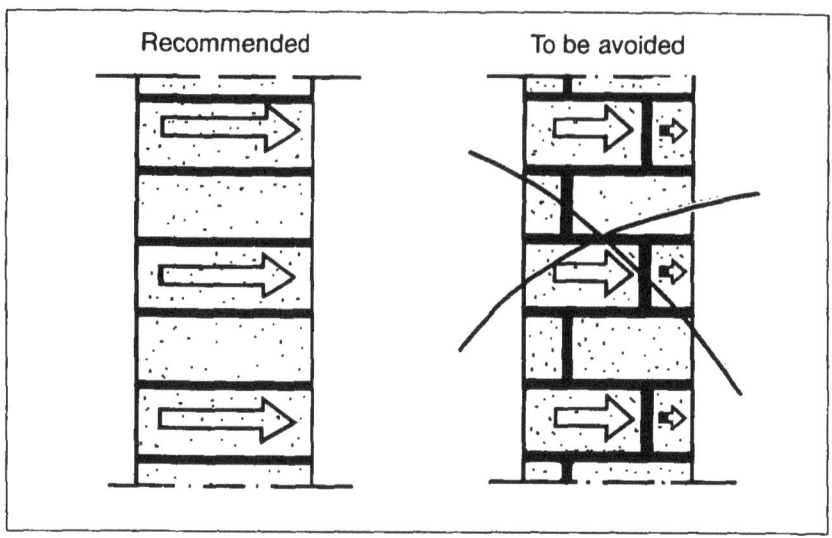

Recommended	To be avoided

Figure 13 Recommended wall construction for absorption and retention of heat

subject to strong thermal stresses. If the earth is exposed and not maintained there will be a drop in the overall performance of the system.

The glazing
This is made up of two parts: the wooden frame and the glass.

The wooden frame
Made of timber and an integral part of the wall, this is the basic structure for putting the glass in position.

The glass
Double glazing is recommended so as to guarantee effective insulation of the system, particularly if no night-time insulation is planned. Installing the glass is quite difficult. A compromise has to be found between adequate sealing to keep out dust and enough play to allow thermal expansion of the glass. Using battens helps to maintain even spacing between the two glass sheets. It will always be possible to dismantle it later for cleaning or for replacing a broken pane.

For ease of supply it is better to choose standard-size glass when designing the system.

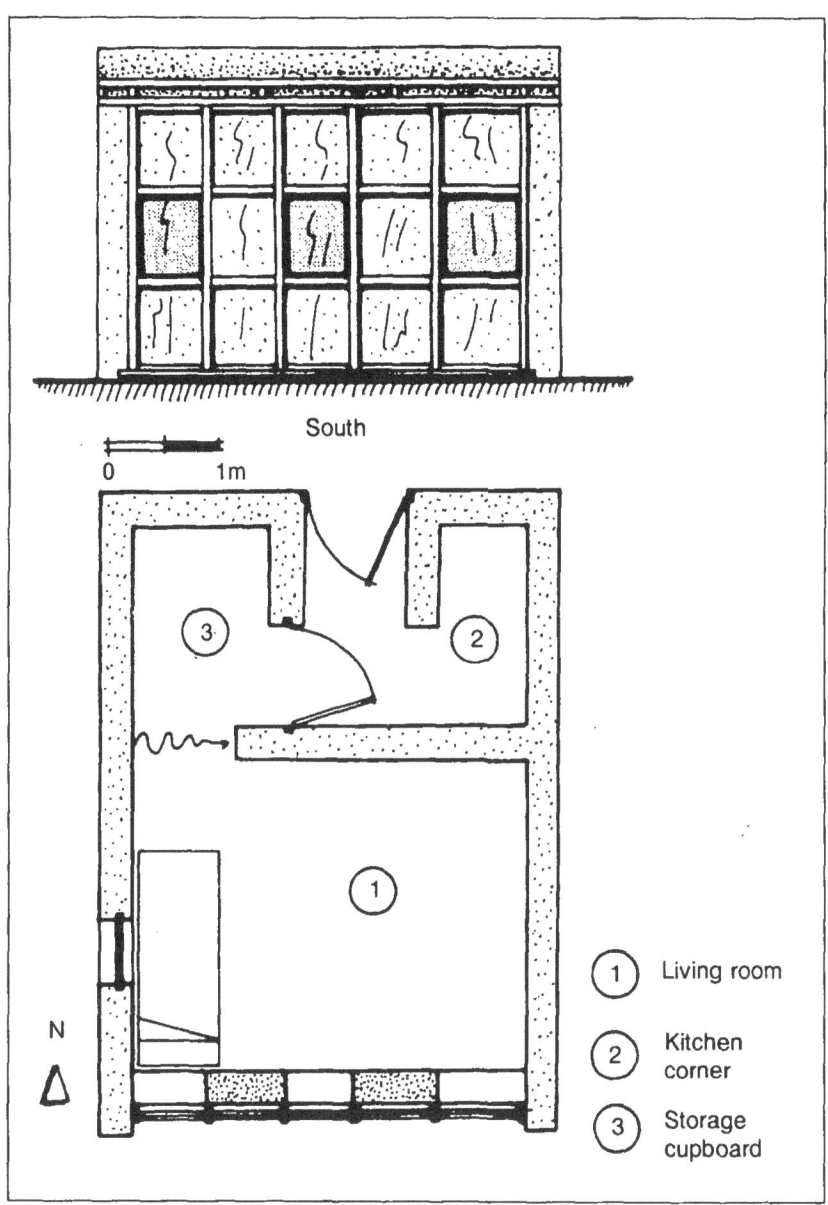

South

0   1m

N

1 Living room

2 Kitchen corner

3 Storage cupboard

Figure 14 Plan of Mrs Spalzon's house studied in Leh, Ladakh

Sizing

Thickness of materials

There is an optimal thickness for each material according to its thermal properties. This ideal thickness increases with the thermal conductivity of the material, that is to say the speed with which it conducts heat.

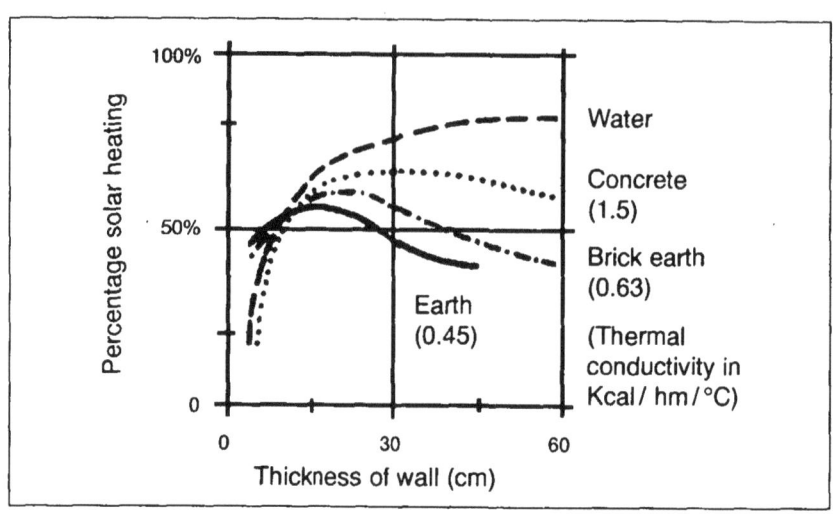

Figure 15 The proportion of heat provided by a solar wall in relation to heating requirement of a building versus the thickness of the wall for various materials
(with acknowledgement to J.D. Balcomb, L.A.S.L., cf. Bibliography)

The more conductive the material, the more quickly it will store energy for a given period of sunshine and the more quickly this energy will be released to the interior. Increasing the thickness increases the quantity of heat stored and prevents it from being released too rapidly to the interior.

A single type of material can have extremely variable properties (see the graph showing variation in thermal conductivity of earth with level of compression, Figure 16). It is therefore difficult to give precise dimensions. However, for cold regions the following rules of thumb can be given:

Material	Recommended thickness (cm)
Earth	25–35
Bricks	30–40
Concrete	35–45

On the other hand an increase in thickness delays release, but also reduces the extent of variation in internal temperature (Figure 17).

For efficient energy release, the solar wall must be three to five metres deep. Greater depth will reduce the efficiency of the system and the level of comfort in the building.

38

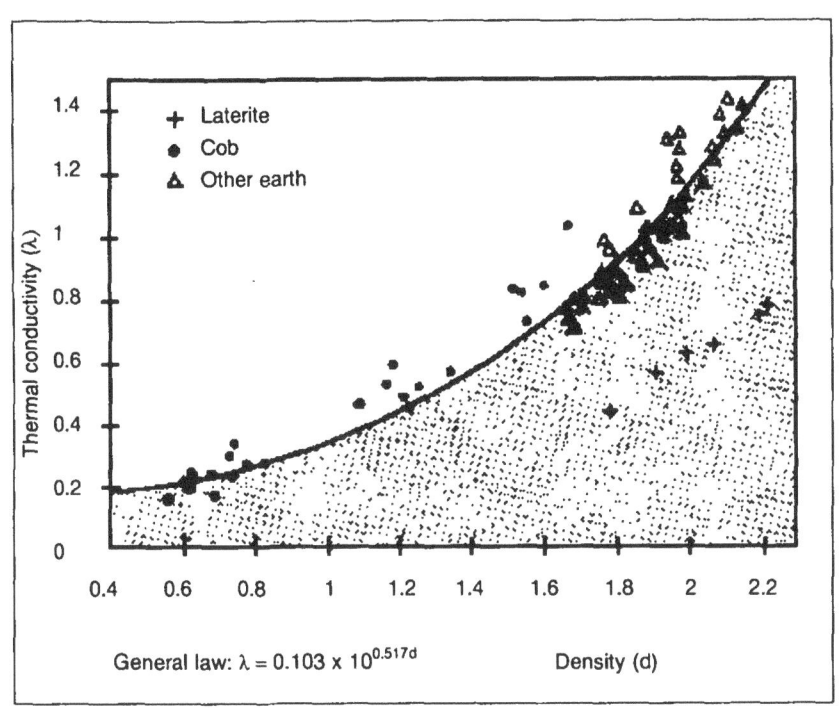

General law: $\lambda = 0.103 \times 10^{0.517d}$ Density (d)

Figure 16 Relationship between thermal conductivity and density (dry material): the more compacted the earth, the more efficient the solar wall
From the thesis by Jean-Paul Laurent, Contribution to the description of the thermal properties of porous granular media (1984).

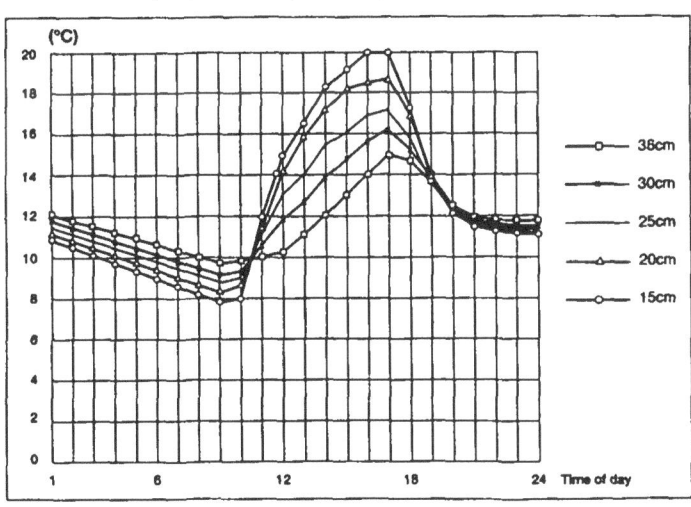

Figure 17 Thickness must be chosen according to desired effect

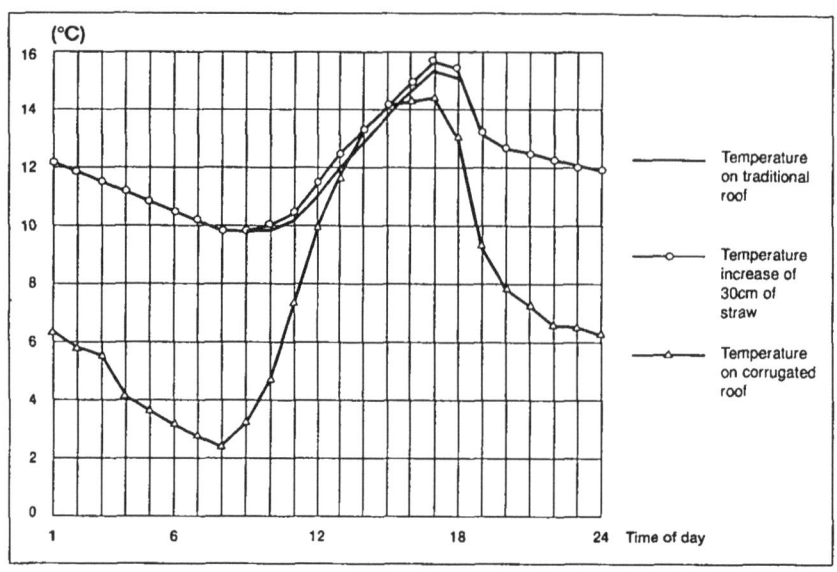

Figure 18 Roofing material is decisive in determining the thermal behaviour of a building; in this case the traditional roofing is sufficiently insulating

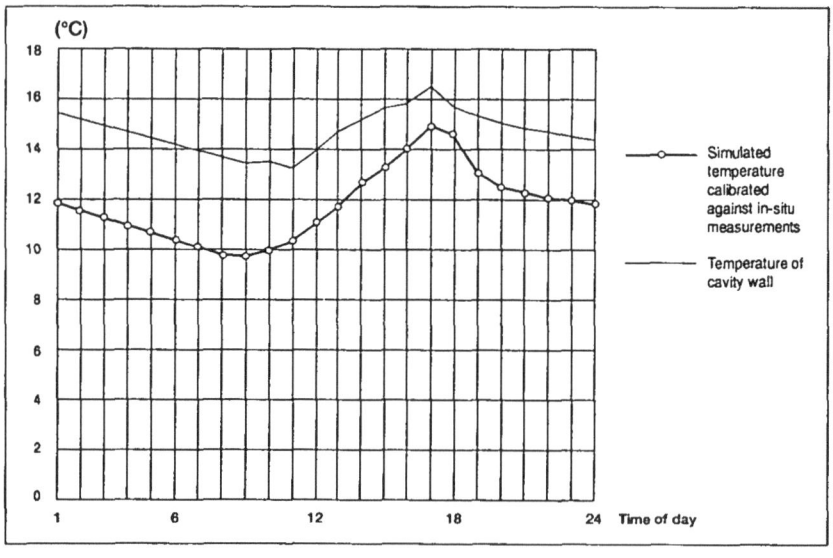

Figure 19 The principle of air cavity insulation remains a simple and effective method of insulation

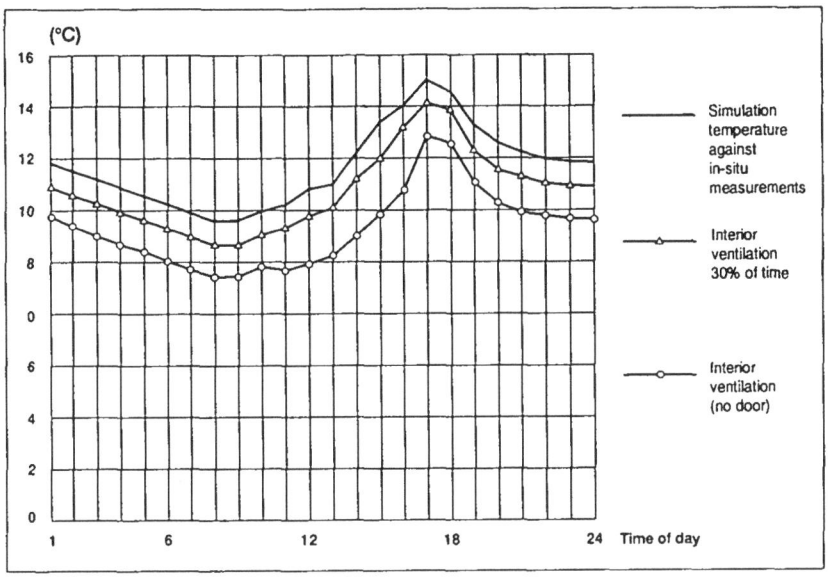

Figure 20 Whatever the heating system, the influence of the inhabitants remains considerable

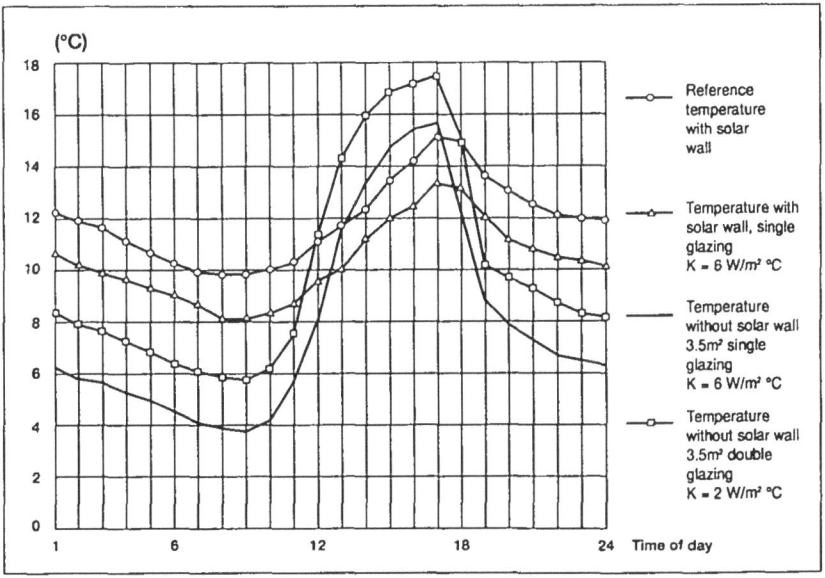

Figure 21 The role of direct gain, coupled with greenhouse-effect wall

41

Thermal insulation of building to be heated

As for any other heating system, the results will be all the better if the building is insulated from the outside climate.

Insulation methods mainly involve traditional roofing (Figure 18) and east and west walls having air cavity insulation within the wall (see Figure 19). There should also be little convective air exchange with the buffer zone (Figure 20).

In Part Three, some simple insulation methods will be described.

Stabilizing effect of greenhouse-effect wall

It is the best system for achieving a small increase in internal temperature (Figure 21).

Details

The design of a collector system on a vertical surface depends heavily on the height of the sun and thus on the latitude of the location concerned. A different system is advised for latitudes between 25° South and 25° North (inclined greenhouse – veranda type for example).

The date 13 February 1993 has been chosen as a representative clear winter day (Figure 21). The influence of the following parameters was tested for this day:

o orientation of the wall
o thickness of the solar wall
o insulation of the building

For greatest efficiency in winter, a wall with a south-facing aspect is recommended in the northern hemisphere and a north-facing aspect in the southern hemisphere, with a maximum deviation of 20°.

Upkeep and maintenance

The timber structure needs protection against humidity and insect attack. A coat of paint or sump oil significantly prolongs the life of the wood.

It is also necessary to check that the system is well sealed to prevent outside air getting in, and broken glass must be replaced as quickly as possible.

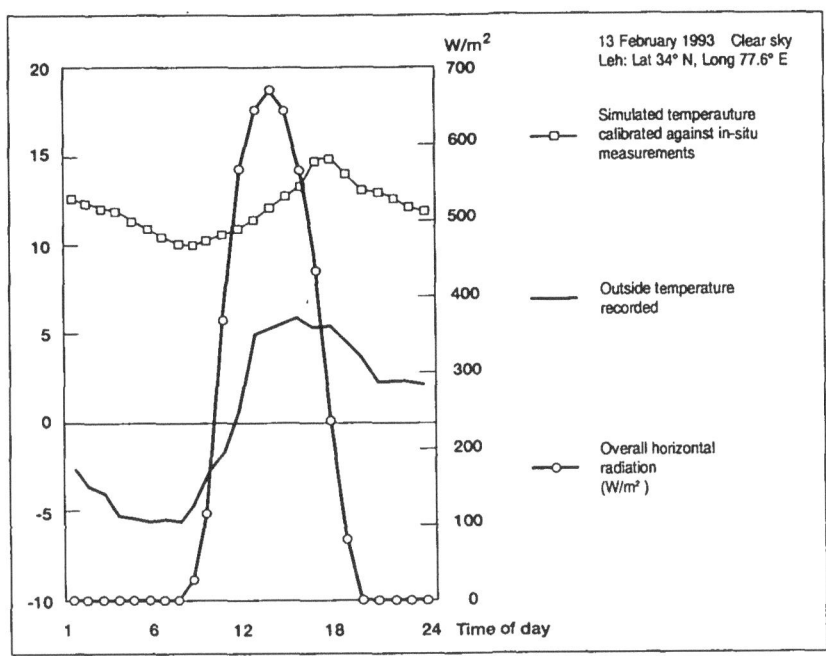

Figure 22 Increase in interior temperature of a house for the sunny day of 13 February 1993; the lag is about 5 hours, which gives highest interior temperature at 18.00 (6 p.m.)

Case 2: The Trombe wall

An example of the Trombe wall system was applied to the library of the Ecological Centre*, based in Ladakh. Thermal monitoring carried out over a winter season makes it possible for performance of the wall to be assessed and comparison to be made with a non-heated room (Figure 23).

Construction principles and usage

The Trombe wall takes its name from its French inventor. It consists of a wall which uses the greenhouse effect, but with ventilation. During the day, the air trapped between the glass and the wall heats up and *thermosiphoning* is established. The lighter, warm air tends to gather higher up.

This layer of warm air is then linked to the interior of the building through two openings made in the inner and upper level of the wall. This allows a *convection current* to flow: the cool air of the building is

---

* Ladakh Ecological Development Group, a Ladakhi association specializing in renewable energy applications, Leh 194 101, Ladakh, Jammu and Kashmir, India.

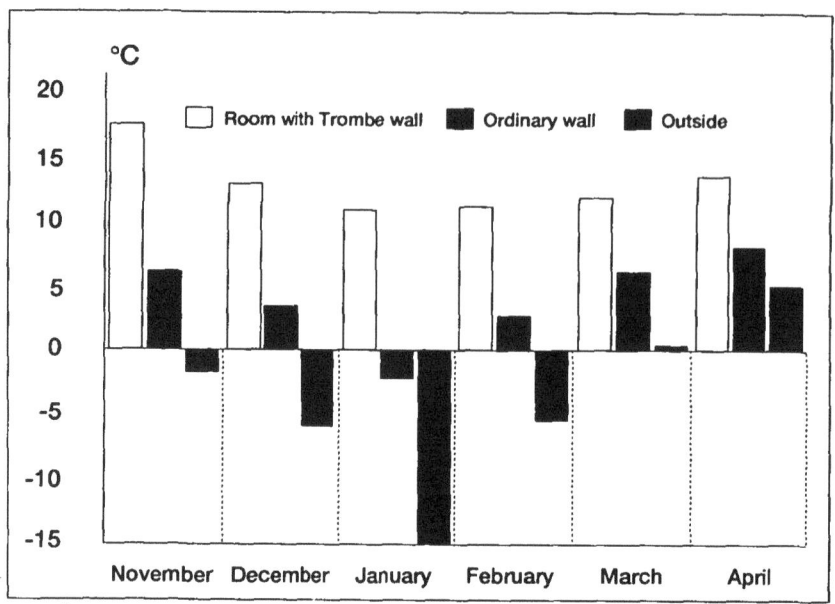

Figure 23 Performance of Trombe wall at the Ecological Centre (1990–91)

drawn into the lower part of the Trombe wall and returned to the upper part of the building, with a significant increase in temperature.

In this way heating can be provided as soon as solar radiation appears, in other words very much earlier than with an unventilated greenhouse-effect wall. An increase in efficiency of around 10 per cent is also obtained as compared to a greenhouse-effect wall. In fact thermosiphoning reduces the temperature of the layer of air, which reduces the wall's thermal losses to the outside.

Installation

The Trombe wall is made in the same way as the greenhouse-effect wall. For thermosiphoning to be effective the sealing must be good, because it should not allow any outside air to enter.

Sizing

The recommended sizing is the same as for the greenhouse-effect wall. However, in the case of the library, the thickness of the Trombe wall was reduced to 19cm. This choice was made because of the clear objective of having maximum heating in mid-afternoon. The perfect insulation of the building means that a comfortable temperature can be maintained.

44

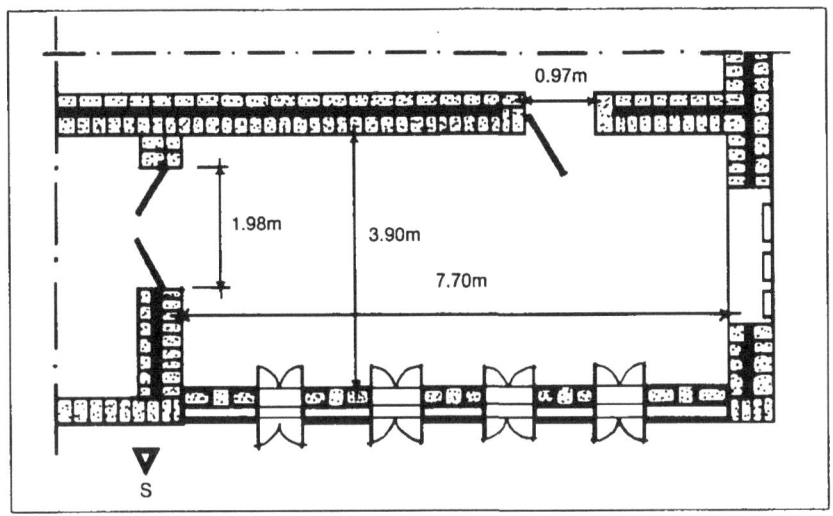

Figure 24 Plan of the Ecological Centre library

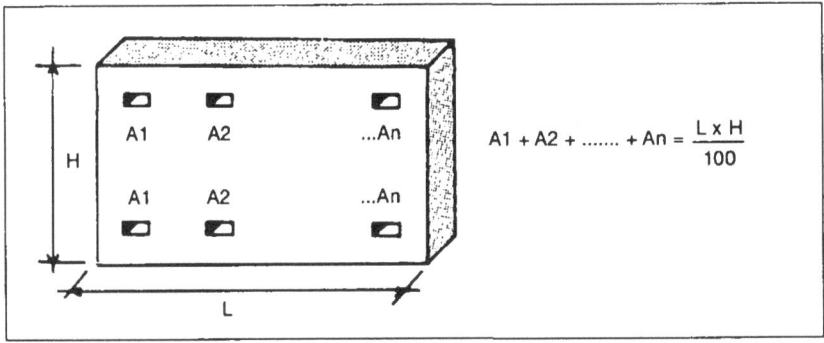

Figure 25 Thermosiphoning openings – surface area
A row of openings of 1/100th of the total surface area is recommended.

Upkeep and maintenance

Unlike the unventilated greenhouse-effect wall, the Trombe wall requires daily operation by the user. In fact, during the night, the natural thermo-siphon movement is reversed. The outside surface of the Trombe wall cools down, and the building becomes the warm area. The warm air is attracted to the upper opening in the wall and will be cooled. It gets heavier and is thus re-injected into the building through the lower opening.

45

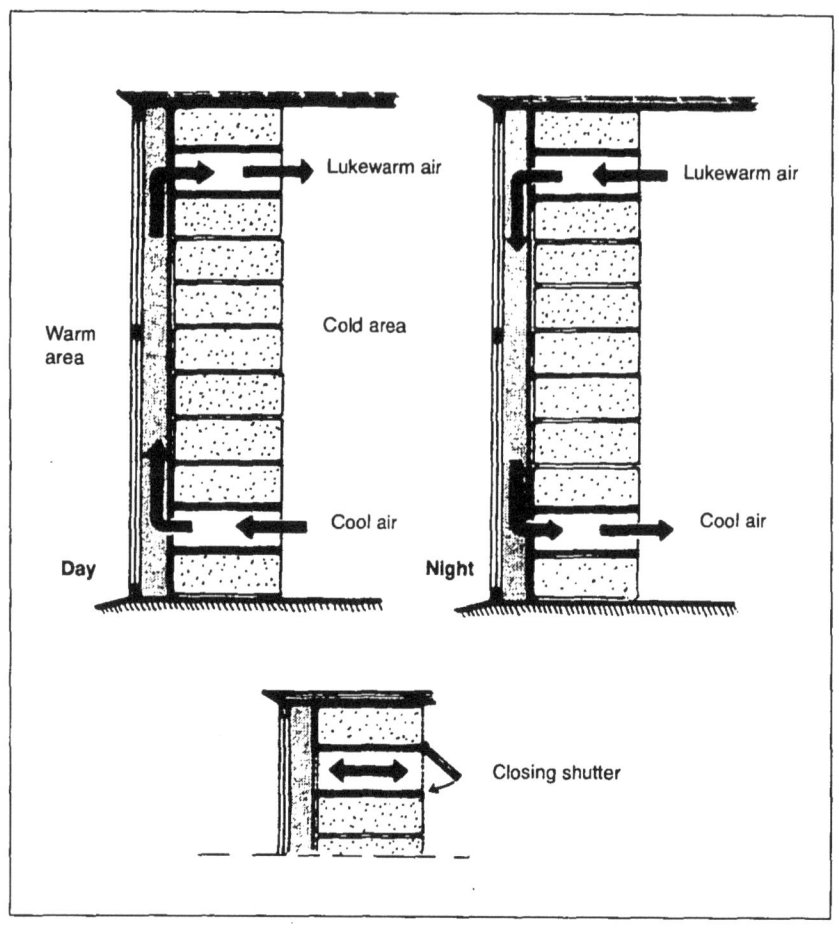

Figure 26 Side elevation of wall

To prevent this unwanted reverse circulation of air the upper openings are blocked off by wooden or metal shutters which can be closed.

There are automatic systems available but they are still expensive. Systems comprising a light covering (animal skin, thin plastic) are too uncertain to be relied upon.

The other significant disadvantage is the dust build-up brought about by air movement between the glass and the outside surface of the wall. It is thus necessary to clean the glass panes once a year, before the winter season. This is delicate work, because it requires the complete dismantling of the glass. Successful cleaning will depend on the skill of the user, who will have to avoid risks of breakage and poor sealing during replacement.

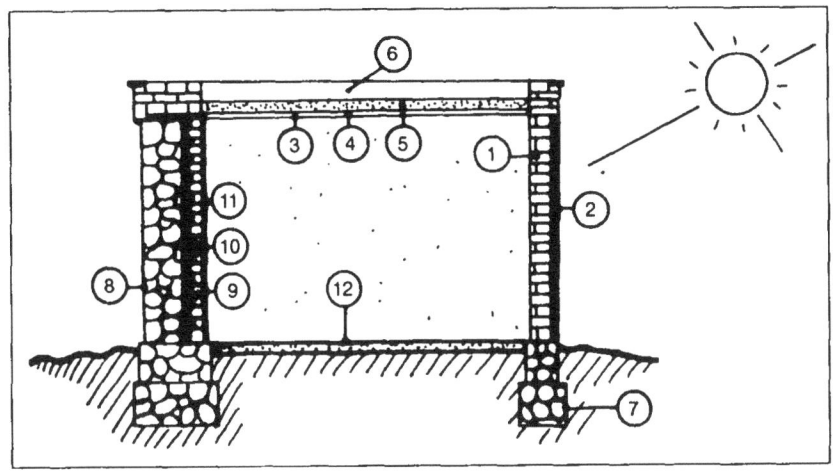

Figure 27 Construction drawing of a proposed building for the Ecological Centre, fitted with a Trombe wall

(1) adobe blocks
(2) double glazing
(3) wooden beam
(4) planks
(5) straw
(6) earth
(7) stones
(8) stones
(9) space filled with straw
(10) adobe blocks
(11) inner coating
(12) concrete or wooden block slab

Case 3: The attached greenhouse – veranda type

Construction principles and usage

The aim is to achieve high temperatures within the greenhouse during the day. The lack of solar radiation at night means that every day there will be wide daily variations in temperature within the greenhouse.

In this process the greenhouse is only considered as a solar collector to heat the attached building. It is never used as a place for growing plants.

The greenhouse attached to housing thus differs from that used for agriculture in two respects: connection with the building and positioning of the main heat storage. The adjacent building to be heated comprises the bulk of the thermal inertia/storage and maintains temperature stability.

For agricultural use, communication is reduced so that the heating needs of the building are prevented from absorbing all the thermal energy of the greenhouse which is required for the plants. Above all, temperature variations must be reduced as far as possible by positioning the main thermal inertia inside the greenhouse.

Installation

The dividing wall

The work of Edward Mazria (see Bibliography) resulted in recommendations of the following wall thicknesses:

Material	Recommended thickness (cm)
Earth	20–30
Bricks	25–35
Concrete	30–45

As in the case of the greenhouse-effect wall, it is important to coat the glazed surface of the wall with a layer of dark colour so as to improve its capacity for absorption of solar radiation (paint etc.).

By adding openings (windows, for example) between the greenhouse and the building, a direct supply of heat can be obtained during the day. Heat exchange is either by radiation – radiation does or does not penetrate the glazed opening – or it is by convection: the warm air from the greenhouse is transferred to the interior through the open windows.

Nevertheless it is important to cover the openings, or to shield the glazed parts during the night in order to insulate the interior of the building because the greenhouse is cooler than the building.

Figures 30, 31, 32 and 33 show the effect of various arrangements on temperatures achieved.

Figure 28 Attached greenhouse veranda

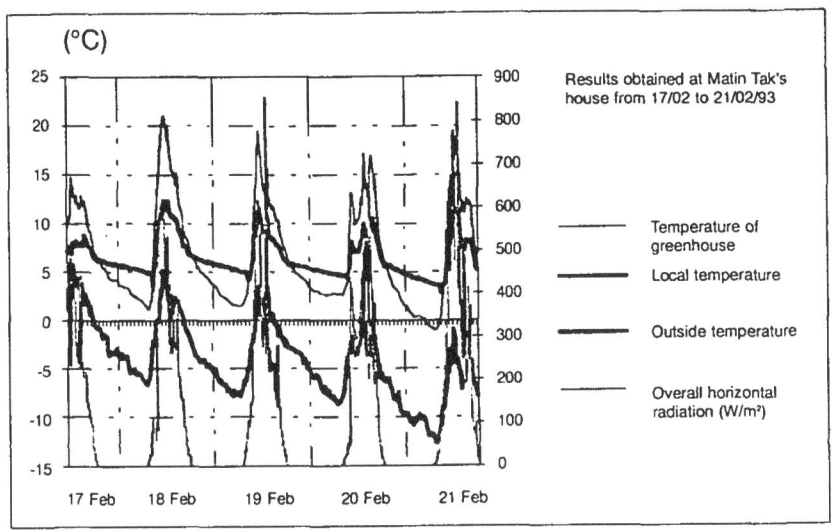

Figure 29 Temperature of greenhouse veranda and attached building
Measured at the home of Matin Tak, Leh, Ladakh, 17–21 February 1993

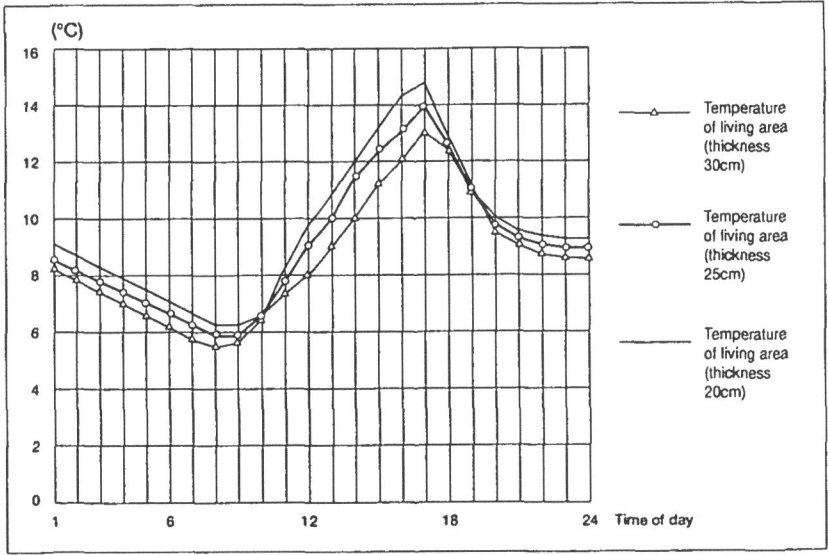

Figure 30 The thickness of the connecting wall determines the magnitude of variations in inside temperature of the adjoining building

49

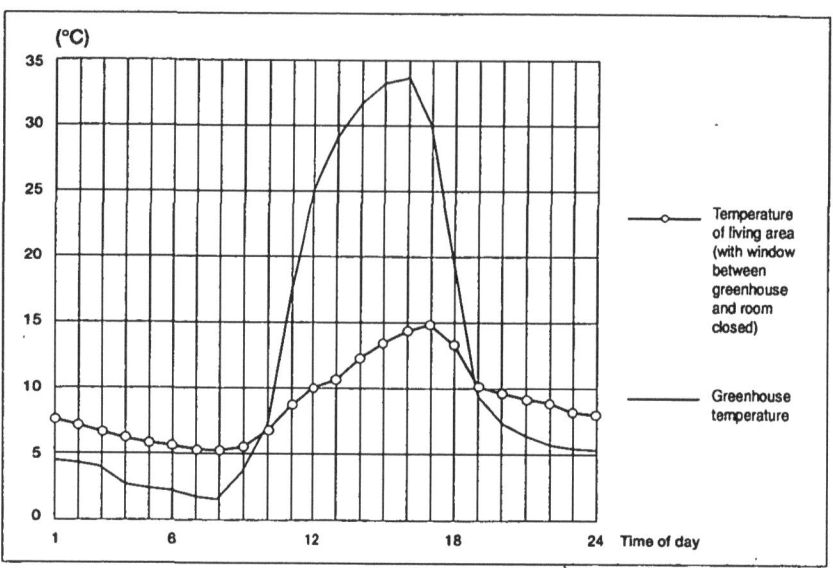

Figure 31 Modelling for 13 February 1993, connecting a greenhouse veranda to Mrs Spalzon's house; temperature variation is greater than that afforded by the solar wall

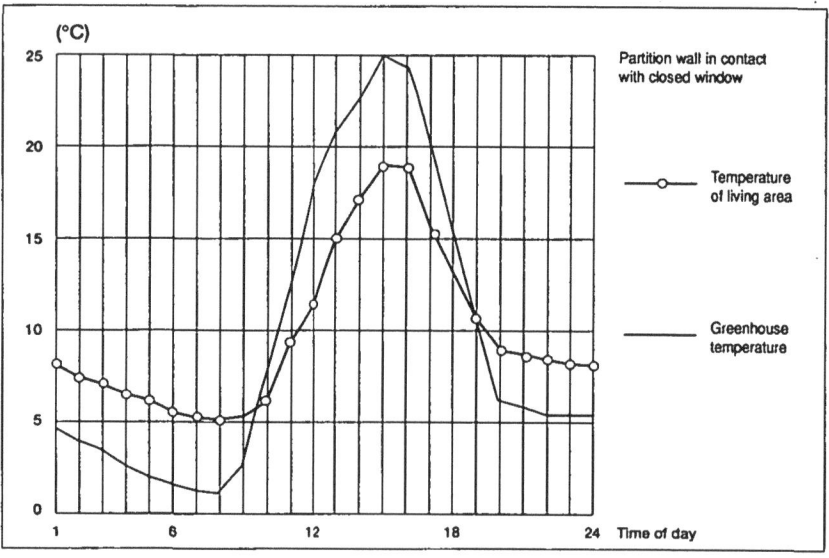

Figure 32 Daytime ventilation between greenhouse and attached living space ensures rise in temperature in the afternoon

50

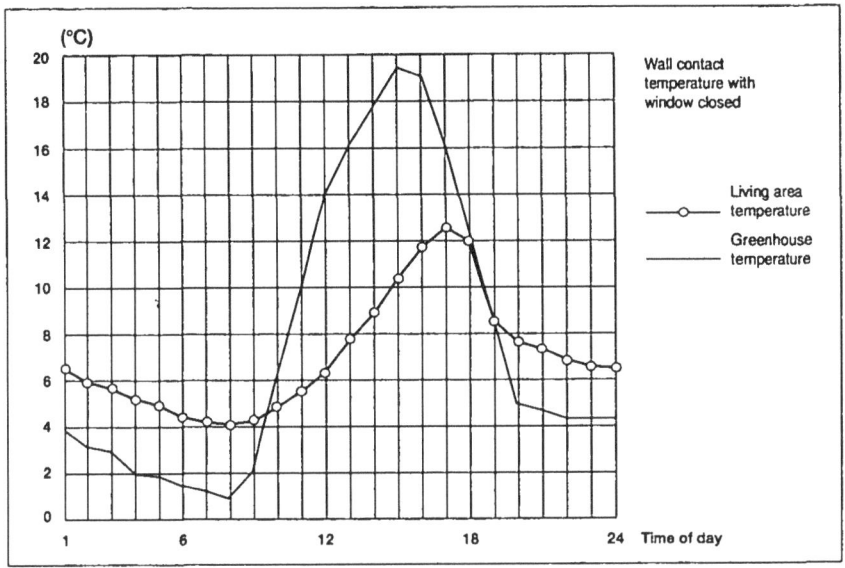

Figure 33 Ventilation by upper and lower openings

The absorptive surface

For the technical elements, refer to the model of the solar greenhouse in full spectrum glass.

For optimal absorption in winter (see monthly solar diagrams in Appendix III) an angle of inclination 'i' must be chosen according to the latitude, where $i =$ latitude $+ 15°$.

The overall structure must be glazed so as to obtain maximum collector area during the day. The best solution is still a roof with a slope 'i' in a glazed frame and a vertical glazed wall, south-facing ($\pm$ 20°) in the northern hemisphere or north-facing ($\pm$ 20°) in the southern hemisphere.

Special features

The principle of the attached greenhouse is universal. Its main advantage is its adaptability to a variety of climatic and geographical situations. Thus for low latitudes where greenhouse-effect walls are less effective, the greenhouse still has a role. It is a comfortable place. During the winter season it is a pleasant place to be in the daytime, and it can become a new living space. In Ladakh, it is customary to put sick people in the greenhouse. If a few plants are placed in it, a higher humidity is guaranteed. This is often pleasant compared with the usually dry atmosphere inside the house.

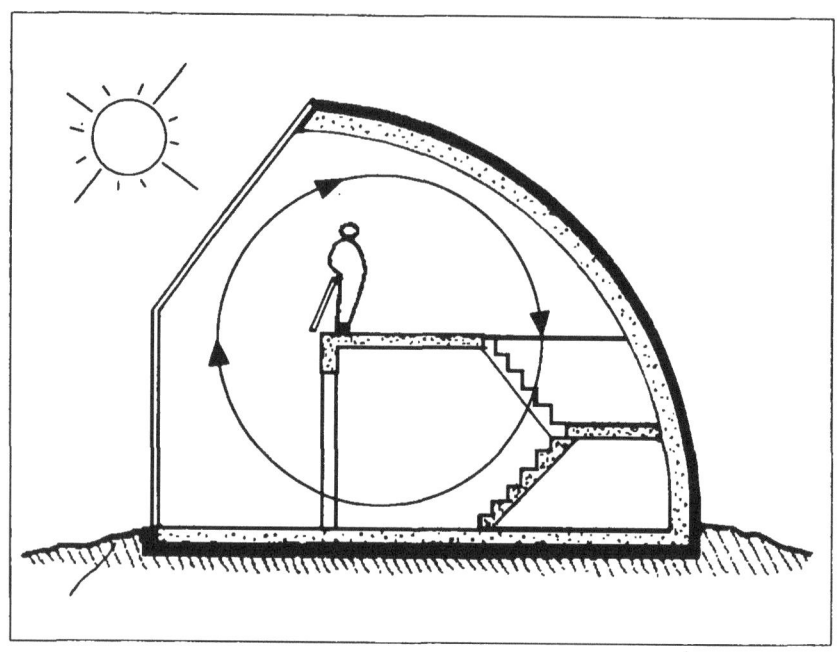

Figure 34 Principle of the convection loop

The greenhouse adapts to different styles of architecture. It is very easy to fit an existing building with a greenhouse. The greenhouse is joined to the connecting wall by deep anchoring with corresponding foundations which provide stability and a perfect seal so that the system is effective.

A model for a single-storey house has been tested by J.D. Balcomb, in New Mexico, USA. The natural convection currents bring about a 'convection loop' which contributes to efficient heat distribution in the upper and lower parts of the building.

Case 4: Mixed *hammam*/solar wall system

This hybrid system was developed in Ladakh to keep a constant high temperature (around 20°C) in buildings used for specific purposes: a hospital operating theatre and a maternity ward.

Construction principles and application

The system combines a solar wall for background heating with complementary heating from the floor, which offers two advantages: the temperature is maintained, even in long periods of cloud; no combustion is involved in the house, which prevents the risk of build-up of gases such as carbon monoxide (CO) or carbon dioxide (CO_2).

A water tank can also be placed over the exhaust gas outlet to provide a supply of hot water.

In addition to the advantages of air quality and even temperature distribution in the operating theatre, the system also has the third advantage of being more economic than a low-efficiency woodstove.

The thermal wall thus guarantees a minimum temperature level (on average 10°C higher than the outside temperature), while the *hammam* system provides the extra heat needed. It can be used continuously or occasionally.

The operating theatre is unique in the whole of Ladakh, and consequently it is in almost constant use. The floor heating works continuously during the winter season. The *hammam* hearth simply has to be lit with a few logs added every morning. In this way there is no delay in re-heating the room. The number of surgical operations in winter has therefore doubled.

Installation

The tiling
Problems of thermal resistance in normal concrete tiling have to be avoided. For this reason, stone tiles with a refractory clay binder are placed on baked brick pillars. This method therefore prevents any thermal deformation from the heat of the flue gases. The recommended thickness is 10cm.

The tiling has a clearance of approximately 30cm from the under-floor. A layer of sand is placed on the lower part in order to insulate the floor. A slight slope ensures constant velocity in the natural convection of hot gases between the entry and exit (under the tiles).

The main difficulty is preventing infiltration of smoke through the gaps. To prevent this, two staggered layers of stones are arranged and the gaps between the stones are reduced to a few millimetres. They are then grouted with a refractory clay mortar.

The hammam *stove*
The stove is situated in the lower part of the north wall. It consists of an open fire, supplied with wood from outside the building.

The combustion chamber
The combustion chamber fulfils two functions: it provides a suitable mixture of air and combustible gas and it maintains the mixture at a temperature which is suitable for efficient combustion.

To satisfy this second function, it is recommended that baked clay

Figure 35 Basic diagram of the *hammam*

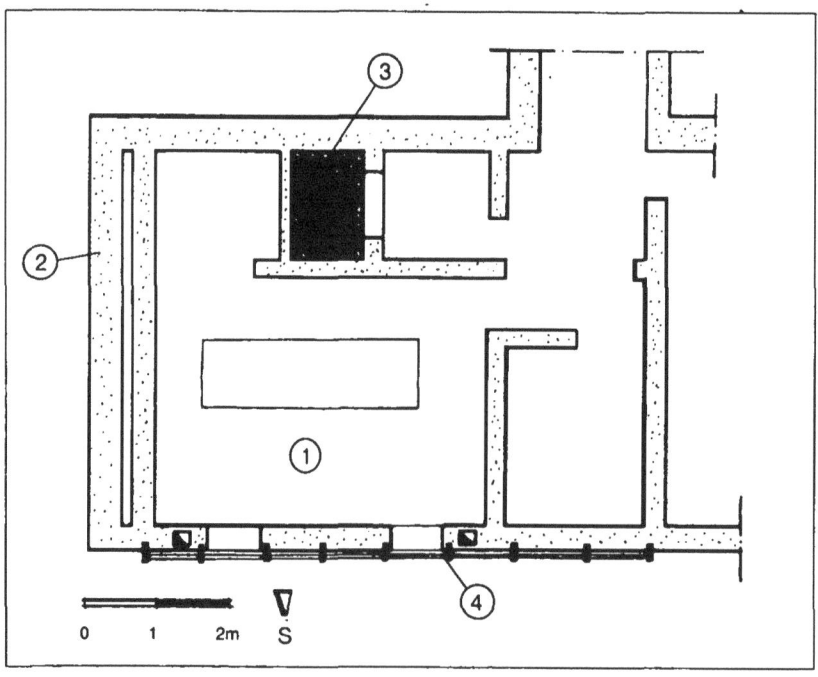

Figure 36 Plan of the *hammam* – *(1) operating theatre, (2) insulated wall (double thickness), (3) position of water tank, (4) solar wall*

54

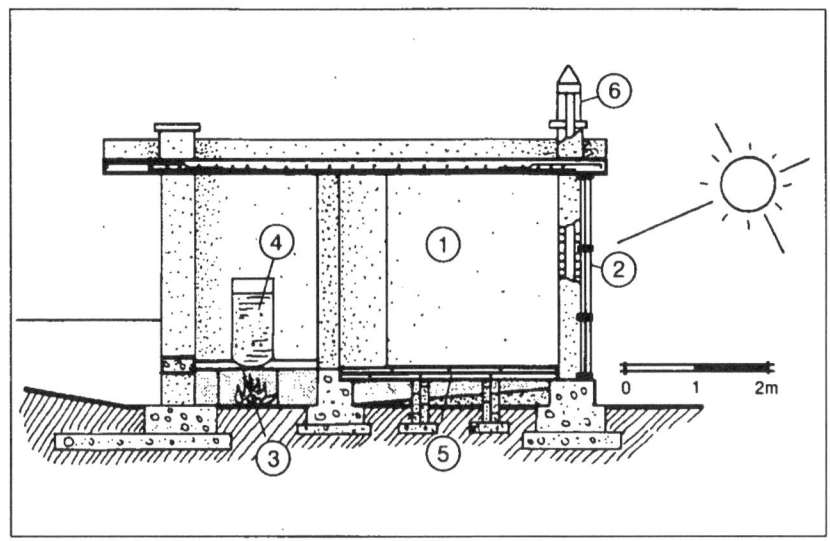

Figure 37 North-south section of *hammam* – *(1) operating theatre, (2) solar wall, (3) woodstove, (4) water tank, (5) under-floor flue gas duct, (6) chimney*

bricks, and if possible refractory bricks, are used in constructing the walls of the chamber.

The walls are then sealed with clay mortar. A vaulted structure proves to be strongest in practice.

It is also preferable to place the wood on a raised grate to obtain optimum combustion efficiency. The air from below passes over the hot embers, reaches a high temperature and encourages combustion. The cinders fall into the lower part thus guaranteeing constant effective heat exchange with incoming air, while avoiding reduction in volume of the combustion chamber.

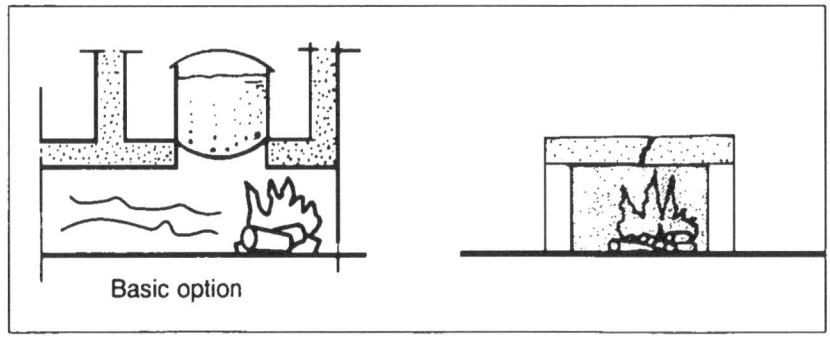

Basic option

Figure 38 Conventional version of *hammam*

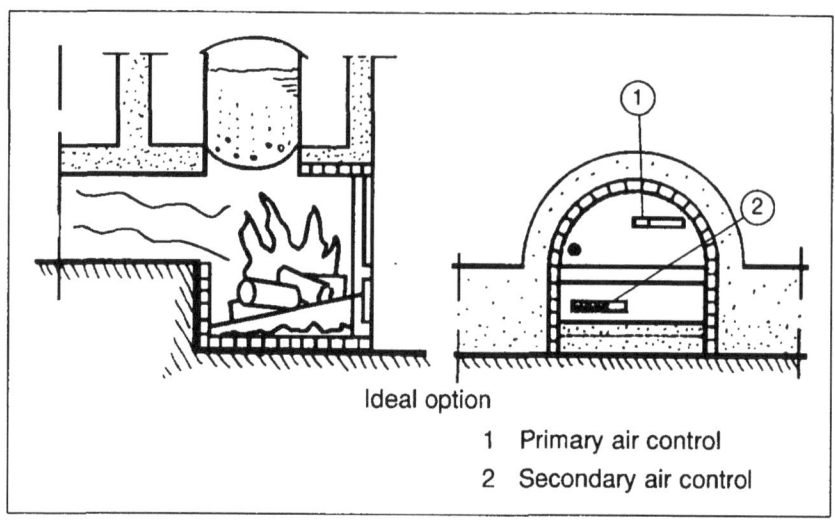

Ideal option

1   Primary air control
2   Secondary air control

Figure 39 Improved version of *hammam*

We present two designs: a conventional but low-performance version, and an improved version.

It is up to the designers to choose, depending on their technical and socio-economic requirements.

There are many disadvantages to the conventional version:

o burst of flames at the start, then complete combustion;
o very poor combustion efficiency;
o no control over the combustion rate, resulting in too high a temperature in the operating theatre;
o excessive wood consumption;
o entry of cool draught on completion of combustion.

With the improved version some parameters have been adjusted to give:

o fine tuning of combustion;
o optimum combustion efficiency;
o cleaning facility by provision of ash container;
o facility for closing chamber completely, or almost completely, when combustion is complete, so as to conserve heat and prevent any fresh air entering.

The chimneys

The flue gases pass through a conduit under the water tank and exit under the operating theatre. Pieces of brick ensure more efficient dispersion by diffusing the gases entering from under the tiling.

On the opposite side, within the thickness of the heat collector/storage

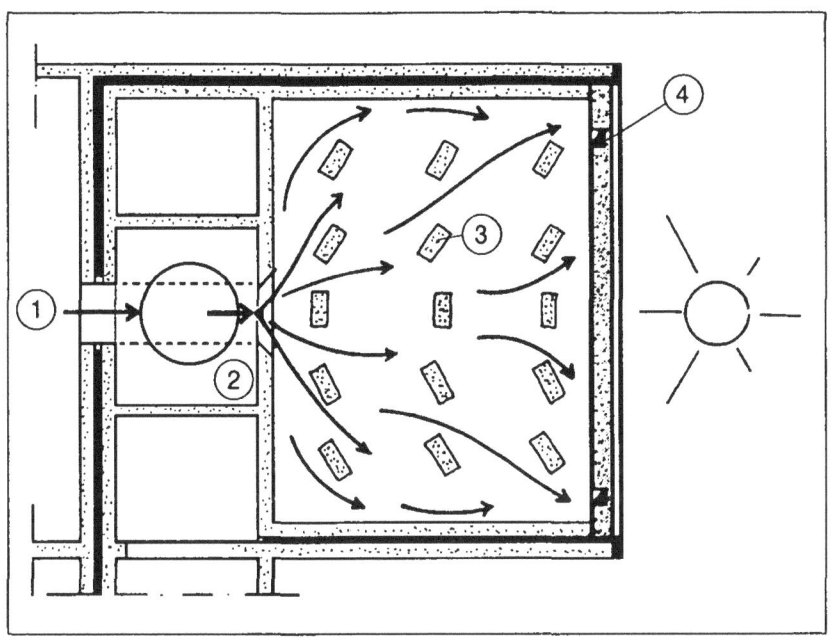

Figure 40 View of *hammam* from above – *(1) combustion gas outlet, (2) duct under water tank, (3) support pillars, (4) chimney*

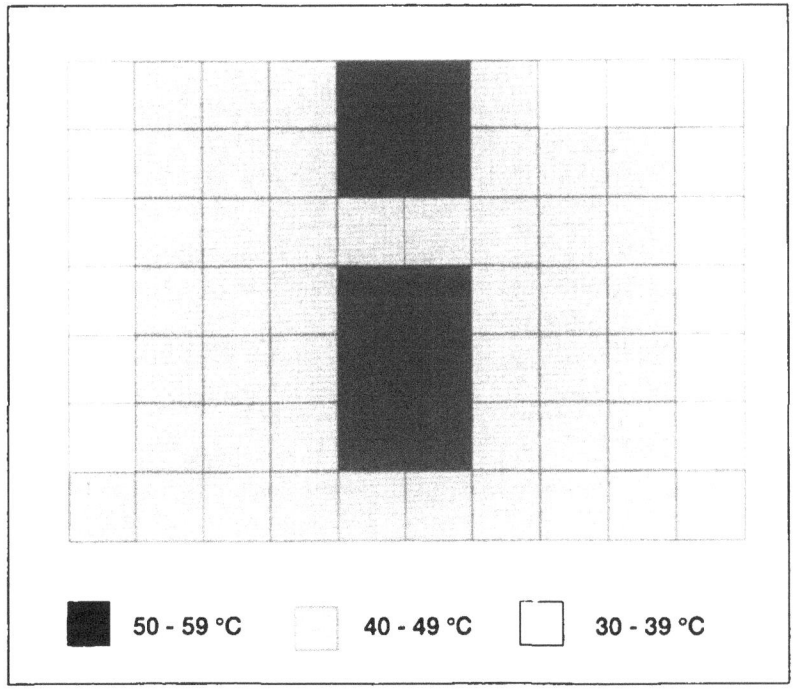

50 - 59 °C   40 - 49 °C   30 - 39 °C

Figure 41 Maximum temperature distribution on the upper surface of the flagstone

57

wall, 30 × 30cm conduits provide even distribution at the level of the slab in the operating theatre.

Dimensions

To avoid too much detail in this section, the reader is referred to Appendix V which provides all details and corresponding calculations.

Maintenance and supervision

The *hammam* process requires regular supervision by an operator. In the traditional version, the following functions could be observed: loading at 7.30 a.m.; monitoring until 9 a.m.; the fire then burning during the afternoon and night until it goes out.

Depending on the daytime temperature, the fuel consumption can reach a maximum of 50kg. This is excessive consumption due to poor combustion efficiency. Furthermore it is difficult to control the inside temperature level with this system. Overheating can result, even when the combustion continues until it burns itself out.

Greater control is vital. It is provided by a single fixed adjustment of incoming air at the start of combustion. This allows adjustment for the varying needs throughout winter.

o In the case of a *very cold day* the first air inlet is opened to its maximum, with a maximum load of fuelwood (50kg).
o For a *sunny cold day*, or a cloudy day which is *less cold*, the air inlet opening is set at medium, as is the wood fuel load.
o On a *mild day*, the air inlet opening is reduced and there is minimal wood loading.

Whatever the pattern of use, dry wood must always be used. Otherwise, part of the heat generated by combustion is used to bring about evaporation of water contained in the wood.

Heat inertia in the tiling is difficult to overcome. It will take four hours from the start of combustion until the desired temperature can be reached on the upper surface of the tiles.

The system of floor heating using hot air from a wood fire therefore has definite advantages: it is efficient; it is hygienic because it avoids the use of particle-laden air or release of CO_2 into the room; and it can be used either continuously or spasmodically.

This high level of comfort in terms of heat and hygiene is particularly well suited to hospital buildings.

4 Examples of agricultural applications

Case 5: The open field solar greenhouse

Construction principles and use

The solar greenhouse is a free-standing, detached structure which can be built away from any other building. It consists of two main parts: a heavy masonry shelter (for insulation and heat storage) and an absorptive surface.

The purpose of such a system is twofold. On the one hand it must provide sufficient heat and light energy in winter to allow market gardening produce to grow; on the other hand, as the seedlings required for planting out are produced in the greenhouse, it should allow early spring harvesting as soon as the outside soil has thawed.

Solar greenhouses have been used for about ten years in Ladakh (altitude 3500m, latitude 34°N, longitude 77.6°E).

Installation

The masonry shelter
The foundations define the perimeter of the greenhouse. They are built of stone to a width of 35–40cm, with 30cm under the soil and 30cm above soil level.

Locally made adobe blocks are chosen for the opaque walls (see Figures 42 and 43). They are about 2.5m high. Earth is used as a binder.

A light-coloured coating is advised for the northern inside wall so as to ensure even light distribution. (Plants placed close to a dark-coloured wall were observed to have much weaker growth.)

The outside wall should also be insulated. This can be done by building it against an embankment and covering it with a mixture of straw and earth. Other kinds of insulating materials can also be added (such as animal wool, vegetable fibre etc.).

Roofing was inspired by the traditional Ladakhi method which involves a supporting structure of sticks from local tree branches which are 5cm thick, a 10cm layer of straw, and a 15cm layer of earth.

A slight slope is preferable, to help disposal of melting snow.

This type of roof is well suited to the dry climate of Ladakh but is also suitable for any other type of sloping roof designed according to other architectural traditions. It is particularly important to ensure that the roof is well insulated. As a general rule a central beam would be installed to prevent the roof from sagging.

In Ladakh, the door is placed at the eastern end. This arrangement can be criticized as it results in poor lighting and inadequate early warming of the interior of the greenhouse. However, this choice is retained in order to prevent the prevailing west winds of the region from blowing in. Less specifically, sufficient glazed area at the eastern end is recommended in order to meet the light and energy requirements of the plants in the early morning.

Regardless of where the door is positioned it must always be insulated.

The collector area
This includes two frames and single glazing.

The frames   For the southern part (in the northern hemisphere) wooden rectangular frames – 1.66 × 5.5m – are used. This is the difficult part of the construction, which requires skilled labour. Joining the two frames, and positioning them at the top and bottom, must be carried out in such a way that they are airtight.

The glazing   The glazing comprises single 3cm glass panes of 75 × 60cm. These panes are placed on square section battens, fixed to the frame, which

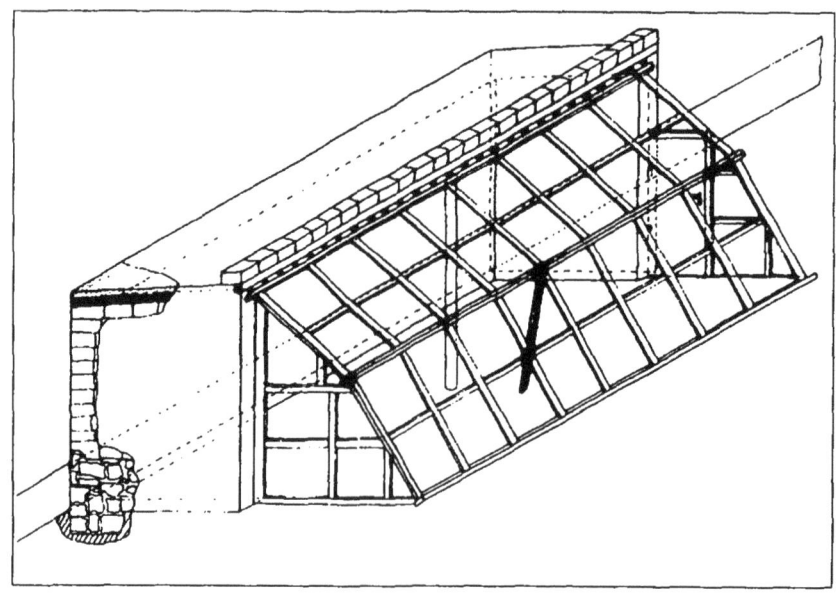

Figure 42 Perspective view of the Ladakh greenhouse

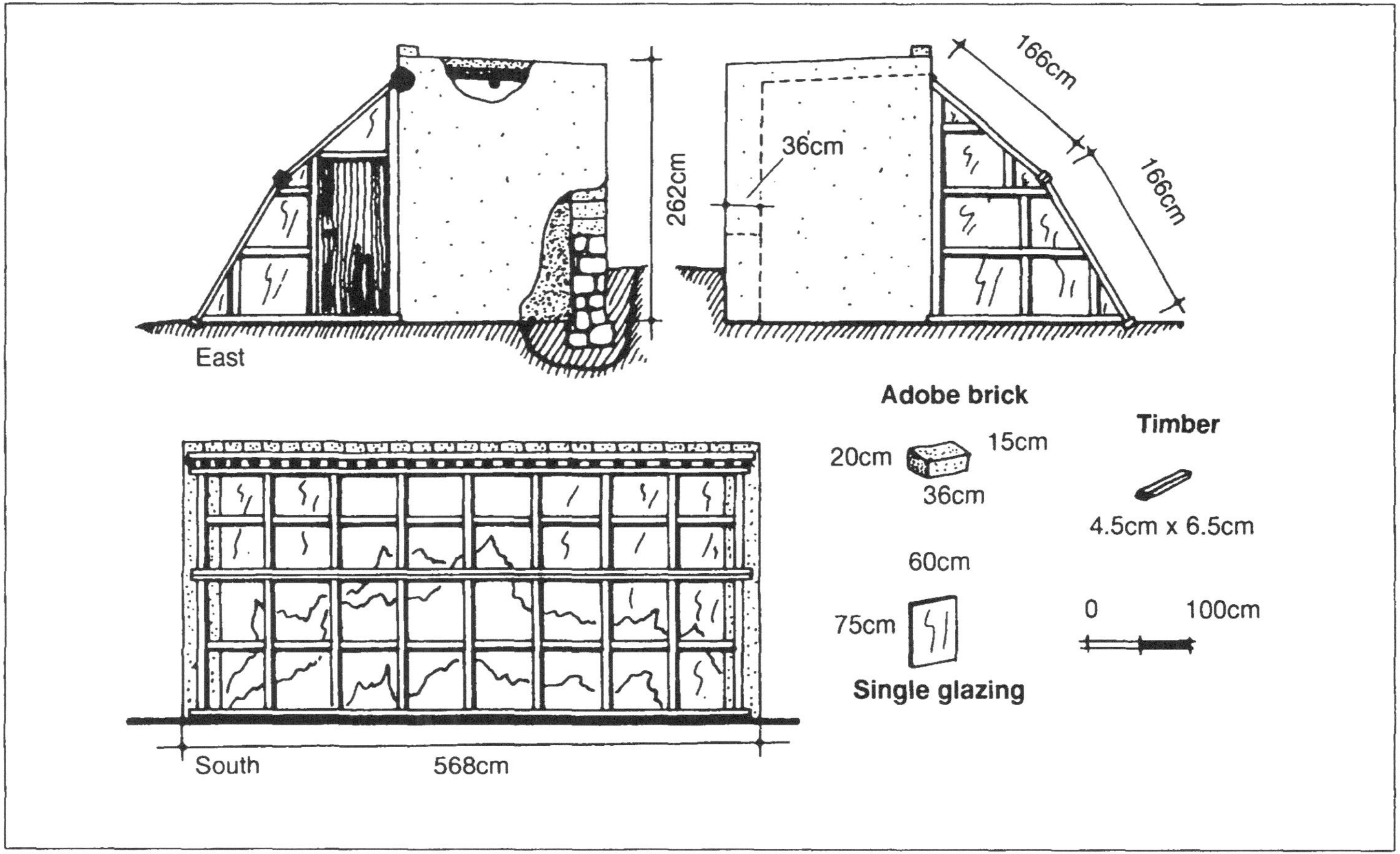

166cm

166cm

166cm

262cm

36cm

Adobe brick

15cm

20cm

36cm

Timber

4.5cm x 6.5cm

0   100cm

60cm

75cm

Single glazing

East

South

568cm

Figure 43 Sketch of Ladakh greenhouse

provides the support framework. The final fixing on the upper part is achieved with triangular section battens.

Once again, the join must be sealed to ensure that the system is perfectly airtight. Mixtures of grease and fine clay can be used although they will require regular upkeep. Insulation of the glazing is still essential, particularly to prevent air escaping (Figure 44).

It should not be forgotten that the collector area needed is a function of the angles of inclination of the upper and lower parts. The former depends on the latitude of the location, and the convention adopted is that the inclination angle of the upper part is latitude plus 15°. The lower part must remain close to the vertical (to catch morning sun, snow reflection).

Sizing

The greenhouse presented here is suited to the needs of a family and the size of the usable agricultural area is around twelve square metres. In this particular case the area was restricted, but larger areas are possible, in which case a balance must be established between collector/storage and insulation, because if the greenhouse is made deeper or wider the existing balance is upset.

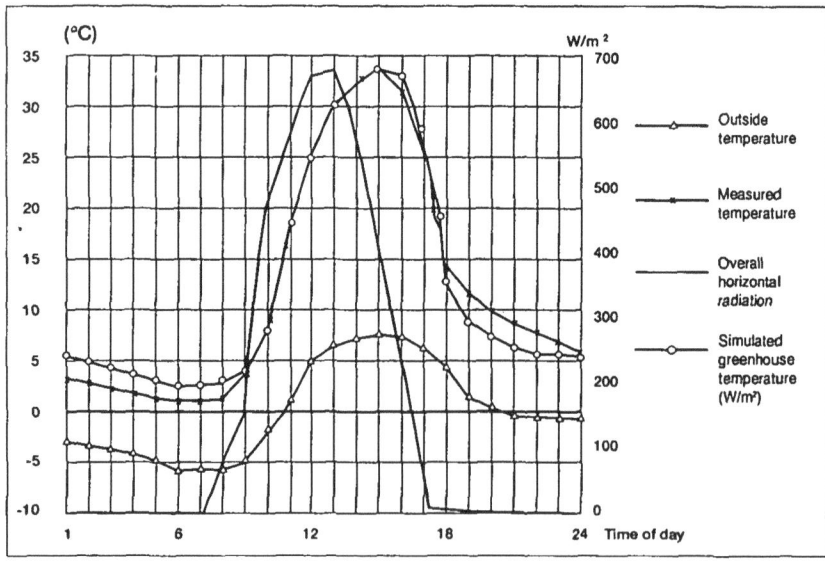

Figure 44 Fluctuations in temperature inside the greenhouse
Temperature was measured on 13 February 1993 and simulated by computer modelling

62

Details

The initial thermal storage is provided principally by the masonry, but intermediary storage is provided preferably by using containers of water.

In general, additional storage is provided by water containers in a proportion of 150 litres per square metre of glazing. This additional capacity must be adjusted in each case until the temperature variations are acceptable.

On very sunny winter days or in summer, there can be a risk of over-heating. To overcome this problem the most effective solution is to provide ventilation via a 'heat valve' during the day, when the temperature is above 25°C. The design must therefore incorporate air holes in the lower part of the northern wall and in the upper part of the east and west walls. These holes remain blocked with cloth-covered straw cushions, for example, when not being used for aeration.

Thought should also be given to reducing excessive temperature variations in any 24-hour period. A rough covering can be placed over the glazed part after sunset.

Finally, the average temperature can be raised, and the daily temperature variation reduced, by adding water containers to increase the heat storage capacity of the greenhouse (Figure 45). The use of a mercury thermometer to monitor interior conditions is therefore recommended.

The last stage consists of matching the thermal inertia to the insulation level.

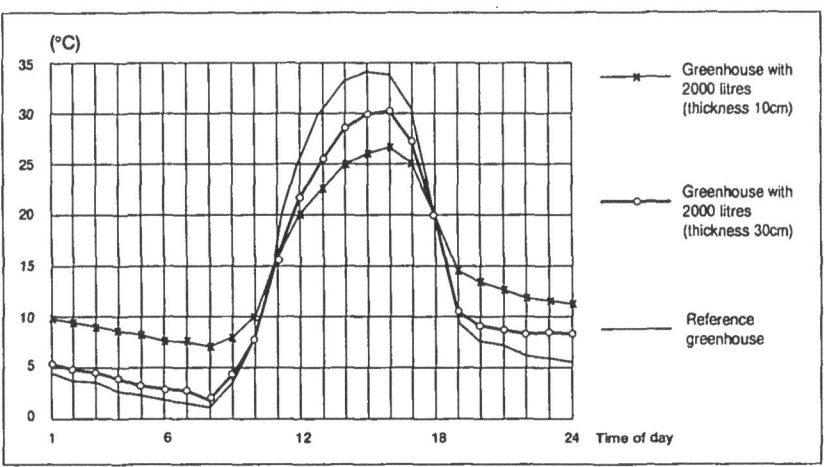

Figure 45 Qualitative effect of the use of water containers for thermal storage

63

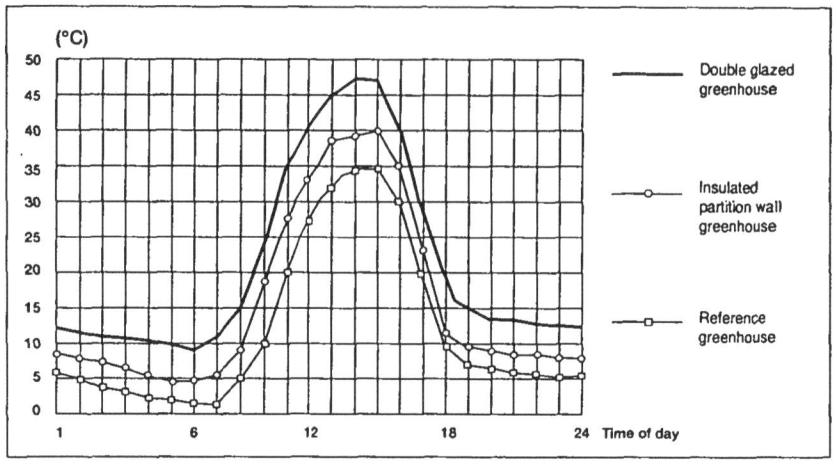

Figure 46 Comparison between insulation of the partitions with double glazing

Maintenance and agricultural operation

To make the timber last significantly longer, particularly in more humid climates than that of Ladakh, it is recommended that the frame of the collector area should be daubed with waste sump oil or paint.

The glazed area of this type of greenhouse requires technical skills not readily available to the user and a significant investment, particularly for non-commercial use.

Payback

The concept of payback period is a basic economic tool for testing return on investment. It consists of knowing how long it will take for the initial sum to be repaid and when revenue will accrue to the owner. For details of the concept of payback time see Appendix IV.

If the produce grown in the greenhouse is intended for commercial use, that is to say if a significant proportion of the produce is to be sold, and the seedlings for planting out are for home consumption, then the payback period is one year.

If the produce is to be used primarily for home consumption, the revenue obtained can balance out the initial investment after four years of operation. We have taken the case of farmers who are remote from markets for perishable vegetables and who do not use planting out techniques. In this case, only part of the traditional winter vegetable production is resold in winter. We have also based our calculations on a current cost of FF2500 (US$420) for a glass greenhouse with 20 square metres of floor area.

For details of agricultural monitoring see Appendix I.

64

Case 6: Polythene greenhouse

To encourage this type of construction the Indian authorities in Ladakh have started a programme based on providing future users with polyethylene film. These greenhouses have consequently developed rapidly in the region.

Construction principles and usage

With this type, the glazed area is replaced by a plastic film (polyethylene) which makes the greenhouse simpler and suitable for self-assembly.

Installation

The absorptive area
The polythene greenhouse differs from the previous model only in terms of the absorptive area. The wooden structure which secures the polyethylene film is made of rough wood sticks available locally. The film is secured at the upper end of the earth walls by adobe blocks.

To enable the plastic covering to be pulled back during the summer and to make it simpler to install in winter, it is made from a single sheet.

The other parts
For a description of the other parts see the example of the glass greenhouse above.

Maintenance and monitoring

With this type of covering the greenhouse becomes very hot and humid on a sunny day. It therefore needs more careful maintenance than the glass greenhouse. Using a thermometer is recommended. A device to provide natural ventilation can also be installed so as to reduce the effects of heat and humidity.

The life expectancy of polyethylene is limited: it varies from two to six years. Rapid ageing is often caused by handling during the summer which causes it to wear.

Regular removal of snow from the horizontal part of the covering should not be neglected when snow falls.

Recommendations for additional thermal storage by means of water containers are exactly the same as those for the glass greenhouse. This is recommended even more strongly because the absence of heavy roofing reduces the initial thermal inertia of the polyethylene greenhouse (see Figure 47).

For the polyethylene greenhouse, which has a larger absorptive area and

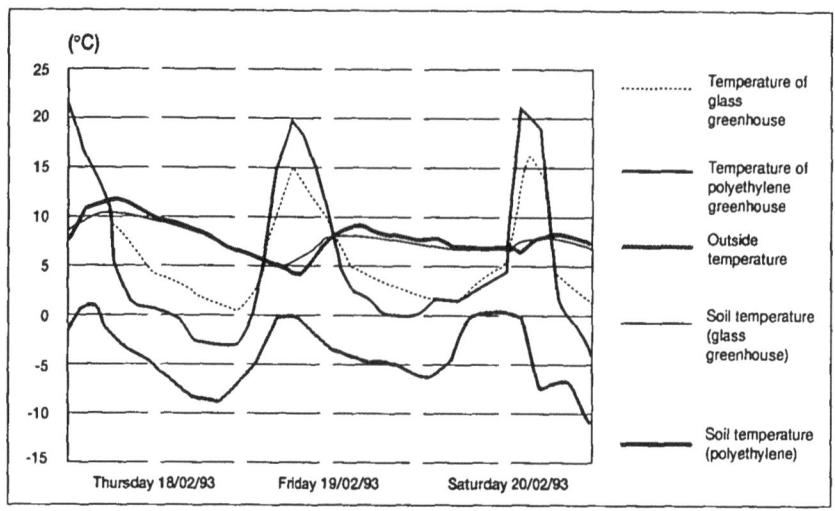

Figure 47 Comparison between the performance of glass and polyethylene greenhouses, carried out in February 1993

less insulation/inertia (the solid roof has been replaced by the polyethylene sheet), greater variation is recorded in the interior temperature. It is thus advisable to retain the solid roof.

Case 7: The attached growers' greenhouse

Construction principles and use

This model is a variation of the open field greenhouse. It offers several advantages:

o cost reduction (no need to construct the north wall of the open field greenhouse);
o close proximity of greenhouse and hence more effective agricultural monitoring;
o use of surplus heat provided by greenhouse to heat the adjoining building.

When existing architecture permits, this solution is advised. However to heat the building, one cannot expect the same results as for the attached veranda-type greenhouse. Here the greenhouse is designed as an autonomous energy system, with its own thermal inertia. Only surplus daytime heat can be recovered. At night the greenhouse acts as an effective buffer zone.

66

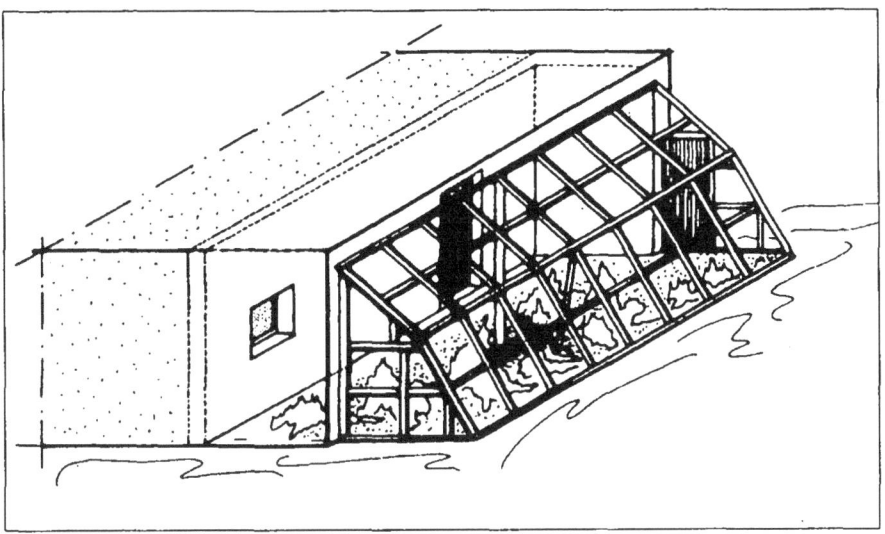

Figure 48 Construction layout of attached growers' greenhouse

This choice can thus be effective for heating an adjoining building, but one must be clear about the objective.

Case 8: Solar hen house for latitudes over 25°

Construction principles and use

In winter, high-altitude regions are isolated for most of the time, as the roads are blocked by snow. It is important for people to make sure that they are self-sufficient in food supply during this period.

The hen house trials using cloches heated with kerosene did not prove cost-effective. Solar energy has thus become not just an appropriate substitute, but also one which can make farming viable.

The proposed model combines two types of solar collector: two roof skylights, which provide direct increase in solar energy, and a greenhouse-effect wall for indirect gain. The system is of course orientated so as to receive maximum sunlight in winter (Figure 49).

Close-fitting covers over the skylights are a good method of keeping out the light in summer, and they have no effect in winter.

Installation

The framework
A timber frame supports the two skylights. Using a covering of earth for the opaque parts is only possible in a dry climate, such as in Ladakh

67

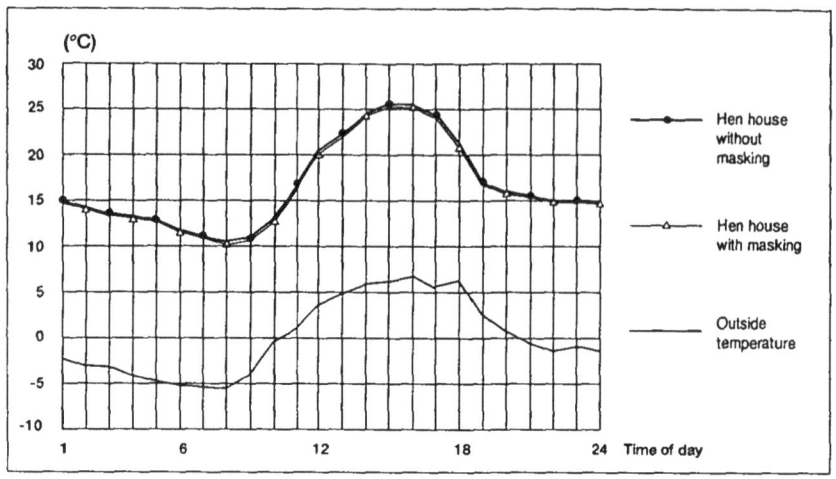

Figure 49 Interior temperature in winter

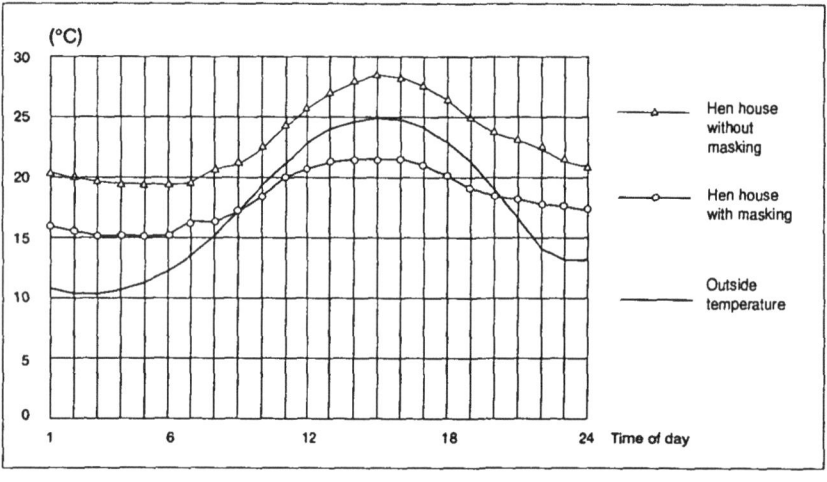

Figure 50 Interior temperature in summer

(latitude 35°N). For rainy conditions it is recommended that a more water-tight covering be used, while at the same time care must be taken to insulate.

Beams or support pillars every three metres prevent the structure from buckling.

Skylights

The orientation angle of the glazing is in accordance with the angle of inclination: i = latitude + 15°, that is 50° in this case (Ladakh). The glazed frames are made by qualified craftsmen and then fitted onto the wooden

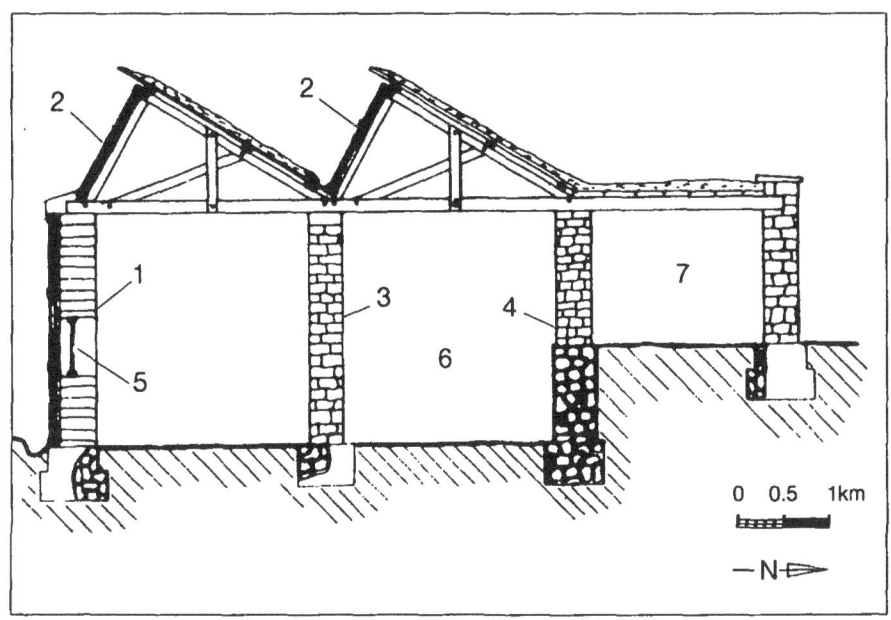

Figure 51 Section drawing of solar hen house
(1) solar wall composed of adobe blocks and double glazing, (2) double-glazed skylights, (3) support pillar or wooden beam, (4) adobe block partition wall, (5) windows for direct solar energy gain and ventilation in summer, (6) hen rearing area, (7) northern access area for storing material and food (buffer zone)

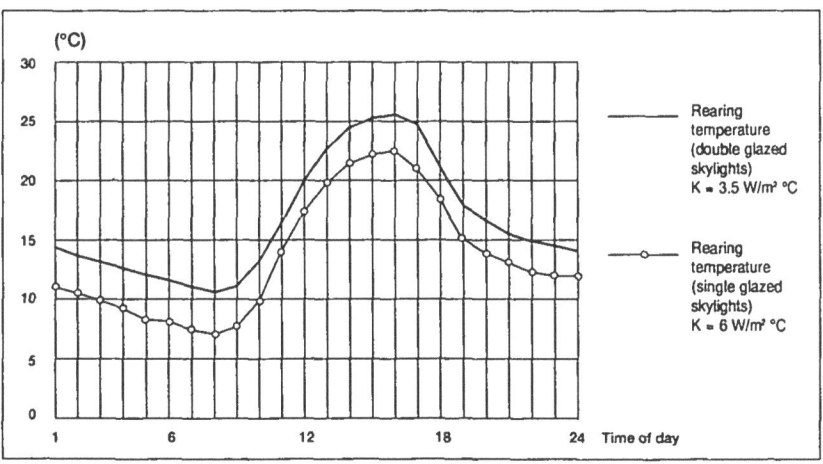

Figure 52 Effect of glazing type (single or double) in skylights

69

structure. A steel sheet covering the upper part ensures that the join between the upper glazing and the frame is watertight. This cover also serves to keep out the sun when it is at its high point during the summer season (Figure 50).

Double glazing is recommended for use in skylights (Figure 52) so as to achieve better overall insulation.

Solar wall

Refer to recommendations given in Case 1 for characteristics of solar wall.

Dimensions

The additional contribution of solar gain from the skylights allows the building to be made five to six metres deep. Obviously the prescribed insulation must be respected and the inertia of the masonry parts (adobe blocks) must be adequate.

The ratio is:

o total absorptive surface/inside volume = $0.5m^2/m^3$
o total absorptive area/inside area = $1.25m^2/m^2$

Details

These systems require special insulation techniques. The whole building is therefore half underground to avoid losses through the east and west sides. On the northern side, the access room acts as a buffer zone and storage area (for food etc.).

The cavity wall technique traps a layer of air in the east and west zones (30cm of adobe blocks + 10cm of air + 30cm of adobe blocks).

It is always possible to add further thermal mass, with black-painted water containers for example.

The surface area of the windows in the solar wall one metre from the ground is equal to one-tenth of the total area of the wall. Opening them in summer provides ventilation and prevents overheating. They must be covered with a grill to stop predators from getting in.

In this system the temperature must not exceed 28°C. Two solutions were suggested to overcome the problems of overheating: a movable covering such as a blanket or cloth can be used to cover the skylight windows, or north-south ventilation can be encouraged using the windows in the solar wall which adjoins the partly raised section of the entrance lobby.

Maintenance and monitoring

This two-source solar heating system can support 35 chicks per m^2 or 15 chickens per m^2.

The basic recommendations for managing rearing of chicks up to adult size are provided in Appendix II.

The dust raised by the chicks dirties the glazed part of the skylights. It is important that they be cleaned before each winter.

Case 9: Low-latitude solar hen house using latent heat

Here we present the model developed by the Runamaqui* Association in the Anccopaccha valley in Peru at an altitude of 3200 metres and a latitude of about 12°S.

Construction principles and use

The intended objective is to maintain the building at a temperature of around 20°C to 25°C for four consecutive weeks, with energy provided entirely from the sun.

This model of hen house combines a structure with high inertia (adobe block walls), reducing the effects of climatic variation, and a collector/ storage system which can take advantage of rapid variations in sunshine from one hour to the next. From the front it has the appearance of a greenhouse, with latent heat elements.

Application of latent heat of fusion offers a number of advantages:

o reduced storage volume for a given thermal capacity;
o storage and discharge at constant temperature: ideally (for pure material) the temperature does not change; in practice the change in phase extends over several degrees.

This type of storage imposes certain constraints:

o cost of material used and its packaging is relatively high;
o the heat exchange area (collector area on the one hand, area in contact with the hot air on the other) limits the storage and discharge phases (the solution is to divide the storage using 0.75-litre paint cans painted black).

Paraffin is also used as a medium of thermal storage. This offers certain advantages:

o paraffin is chemically inert and can be stored in almost any type of container (however, copper and polypropylene should be avoided);
o the fusion temperature can be selected thanks to a fairly wide range of products. Correct mixtures can easily be found (not using iso-paraffins

---

* According to documents provided by Runamaqui authors: Christine Bénard, Dominique Gobin, Béatrice Guerrier, Frédéric Michaux

71

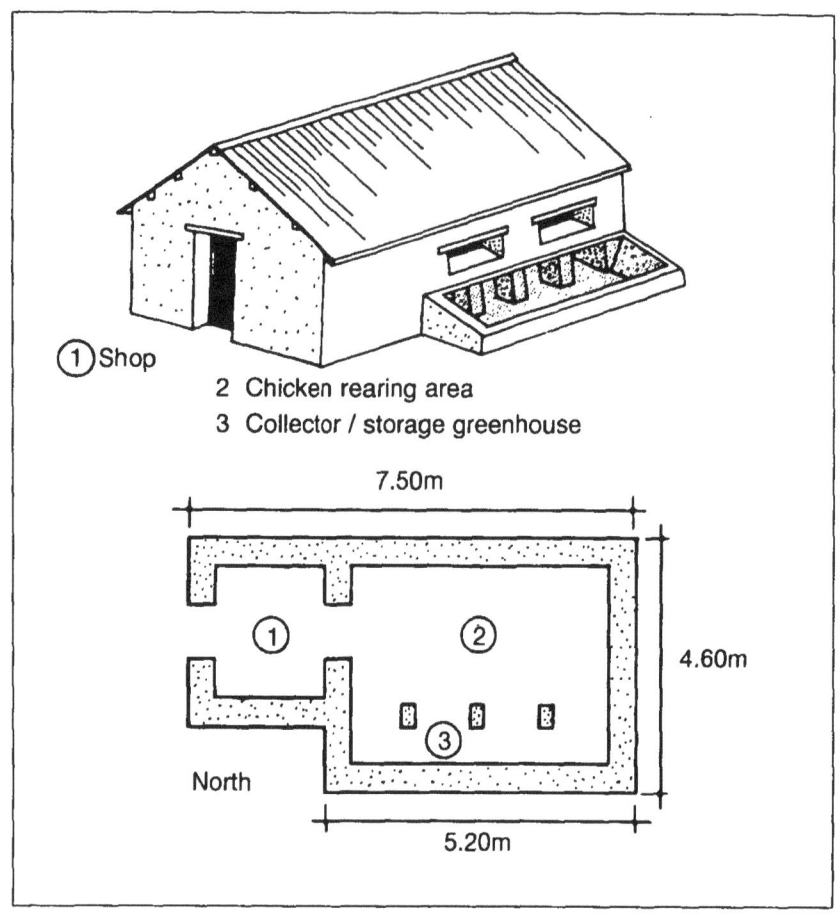

Figure 53 The Anccopaccha hen rearing building in Peru

and especially cyclo-paraffins). These go from 44°C to 62°C, the most common ones being 50/52°C and 58/60°C.

Paraffin use is not, however, without disadvantages:

o often it is not locally available and must be imported from better supplied regions, such as the industrial towns on the Peruvian coast;

o it is also an inflammable product, even if it does not spontaneously combust at normal temperatures;

o it is a poor heat conductor. The thermal resistance of the liquid formed on the surface on melting (or of the solid crust on solidification) must be reduced by encouraging natural convection effects in the liquid or by introducing metallic parts in the medium (iron shavings and filings).

72

Sizing

The Anccopaccha hen house comprises an east-west rearing room, about 4m wide and 2.5m deep. Over the whole width of the northern side a space approximately 1m deep is attached, containing the part which collects and stores solar energy. On the west side, a storage room acts as a screen to prevailing winds and is used as an air lock for entering the room.

The hen house can hold 500 chicks aged from one day to one month.

Details

The architectural components are designed to encourage the use of local materials and construction techniques. This principle does, however, have some drawbacks:

o industrial paraffin (melting temperature 58/60°C, feed quality) is used as an energy storage medium;
o glass is inserted for the greenhouse to absorb solar energy (traditional rural houses do not use glass);
o the roofing is of fibre-cement for reasons of hygiene. This particularly helps with disinfection.

These non-traditional elements can thus be used, but they significantly increase construction costs (cost in 1982 for a hen house for 1000 chicks: US$9000).[*]

Maintenance and upkeep

During chicken rearing, smooth running of the hen house requires:

o all entry and exit doors to be kept closed to keep the heat in;
o insulating shutters which cover the greenhouse during the night to be opened and closed morning and night. This reduces heat losses to the exterior. These shutters have caused some construction problems. After several fruitless attempts with local materials, synthetic insulation (polystyrene) was used as a last resort;
o internal shutters separating the collector/storage space and the actual hen house to be opened and closed morning and evening. The idea is to encourage maximum heating of the store during the day and to heat the building by natural ventilation at night.

For operation of hen house see Appendix II.

---

[*] The outlined plans are available from Runamaqui (cf. Bibliography)

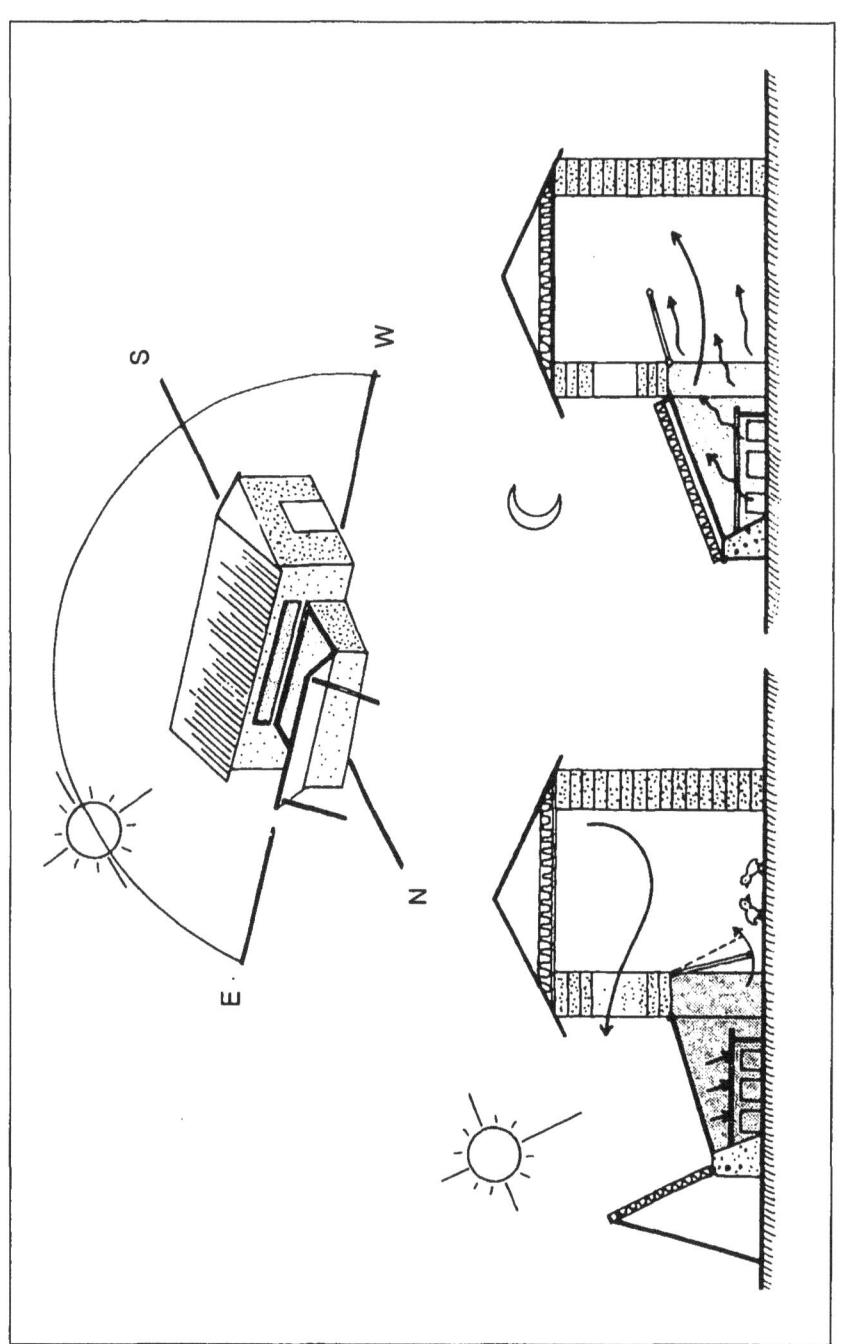

Figure 54 Orientation of the Anccopaccha hen house

PART THREE

THE PHYSICS OF SOLAR HEATING

The aim of Part Three is to introduce the basic theories necessary to understand, design and predict thermal performance of a solar-heated building. It is intended to be an instructive synthesis of theoretical ideas to give a full explanation of thermal behaviour in passive solar systems.

To denote the energy source, the sun, the term 'solar resource' is used. It illustrates the wide variation of its availability on the earth's surface, in terms of time and space.

A given location can be described in terms of the level and consistency of its 'solar resource', as in the case of a fossil fuel (oil, for example). There is however one major difference: solar energy is inexhaustible.

5 The solar resource

The origin of the sun and solar radiation

The sun came into being following the contraction of a huge cloud of gas, consisting essentially of hydrogen, under the force of gravity.

Violent collisions between hydrogen particles then released intense heat, allowing fusion of these hydrogen nuclei with the associated nuclear energy. The current rate of this fusion is 3.83 million tonnes per second. There will therefore be an end to this so-called renewable energy source, but it will still be in abundant supply in five or six billion years.

Thermonuclear fusion releases energy in the form of waves or high frequency electromagnetic radiation. This radiant energy is generated at the heart of the sun at temperatures estimated as being between 10 to 14 million degrees Celsius (°C). At the sun's surface the temperatures are no more than 5500°C.

Then there are photons, described as particles of light travelling through space in the form of electromagnetic energy, which comprise radiation of different wavelengths.

All this electromagnetic radiation is emitted from the sun travelling in all directions at the speed of light. At 150 million kilometres, the earth intercepts 0.45 of a billionth of the power given off by the sun. The intensity of the radiation or of energy radiation reaching the upper layers of the earth's atmosphere is called the *solar constant*. It is the power per unit of surface area, whose value is 1367 W/m^2 (Watts per square metre).

Even though solar radiation comprises all wavelengths, the energy emitted is mainly contained in the visible and near infra-red wavelengths. With a temperature of 5500°C at the sun's surface, this energy is mostly found at high frequencies (short wavelengths).

It is normal practice to separate the spectrum into three parts:

Visible spectrum   This is the band of wavelengths to which the human eye is sensitive. It represents 46 per cent of the total energy radiated by the sun. The wavelengths are found between 0.35 and 0.75 microns (1 micron = 1 millionth of a metre). Violet is the shortest wavelength (0.35 microns), red is the longest (0.75 microns).

77

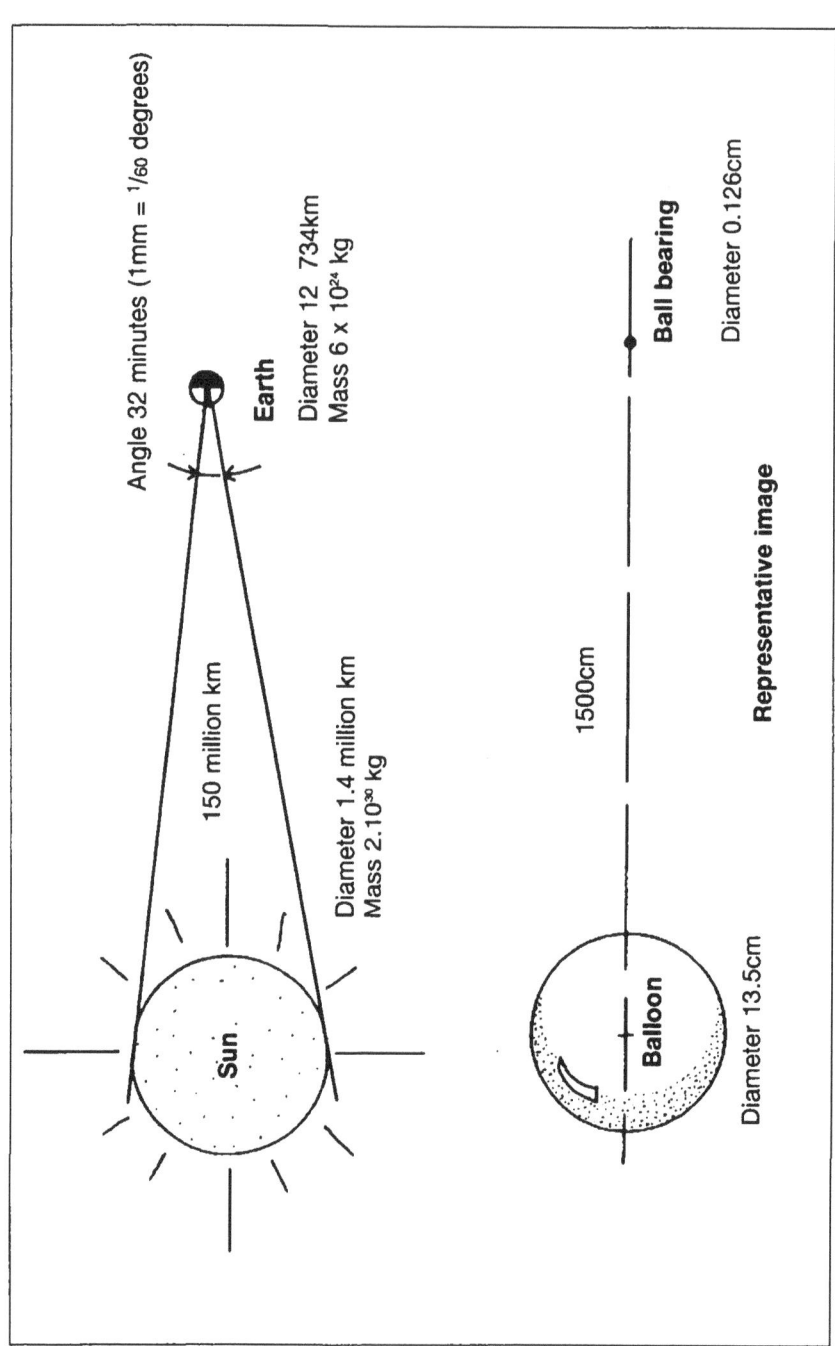

Figure 55 The proportional sizes of the sun and the earth

Angle 32 minutes (1mm = $^1/_{60}$ degrees)

Earth

Diameter 12 734km
Mass 6 x 10²⁴ kg

150 million km

Diameter 1.4 million km
Mass 2.10³⁰ kg

Sun

1500cm

Balloon

Diameter 13.5cm

Ball bearing

Diameter 0.126cm

Representative image

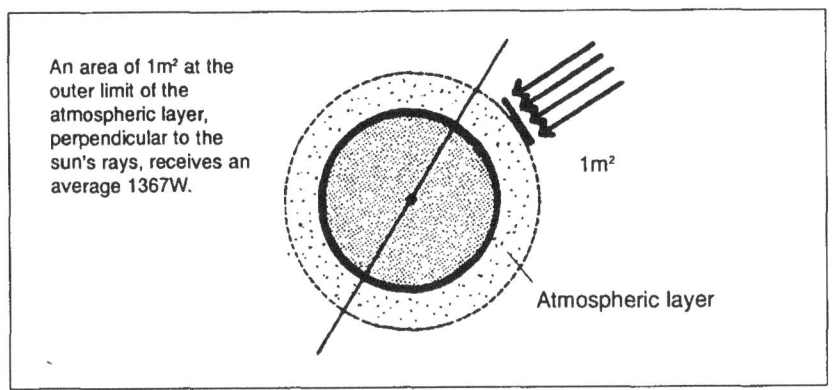

An area of 1m² at the outer limit of the atmospheric layer, perpendicular to the sun's rays, receives an average 1367W.

1m²

Atmospheric layer

Figure 56 Definition of the solar constant

Infra-red spectrum This describes wavelengths over 0.75 microns and represents 49 per cent of the energy radiated by the sun. It is felt as a wave of heat.

Ultraviolet spectrum These are all wavelengths below 0.35 microns.

After the arrival of solar radiation, the next stage is penetration of the atmosphere. Three physical phenomena reduce the quantity of energy which arrives directly onto the earth's surface.

o *Reflection back into space*
  This is due to the presence of clouds and dust in the atmosphere and on the surface of the earth. It can be water, snow, sand etc. The name given to the power of reflection of radiating energy from one of these bodies is *albedo*. Reflection accounts for 35 per cent of the radiation which reaches the earth's atmosphere.

o *Diffusion while travelling through the atmosphere*
  Part of the direct radiation is transformed into radiation diffused in all directions. This is caused by the presence of molecules of air, aerosols and other particles of dust. This transformation is responsible among other things for the blue colour of a clear sky, which diffuses mainly blue from the visible spectrum.

o *Absorption*
  In the upper layers of the atmosphere, ozone eliminates ultraviolet radiation by absorption. Then carbon dioxide and water vapour absorb another part of the radiation. Total absorption represents 10 to 15 per cent of radiation which reaches the atmosphere.

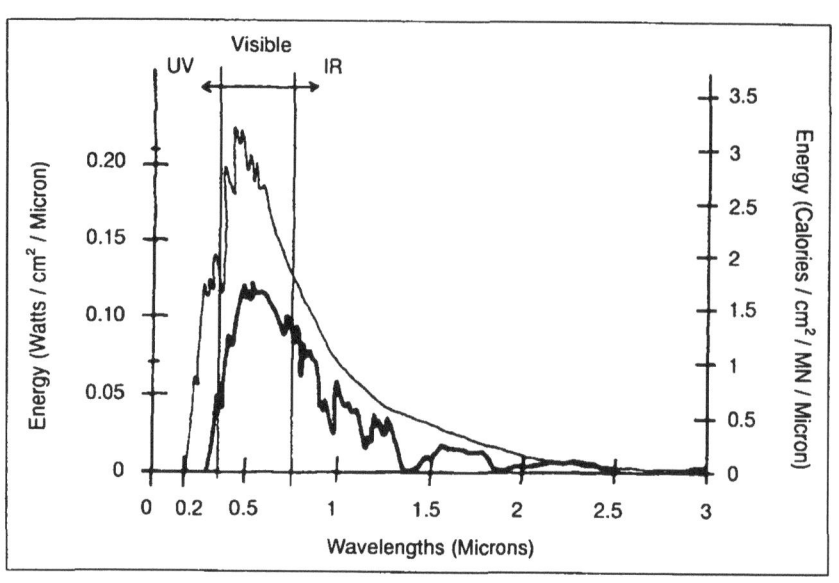

Figure 57 Energy distribution according to different wavelengths at the atmospheric layer (fine line) and the earth's surface (thick line)

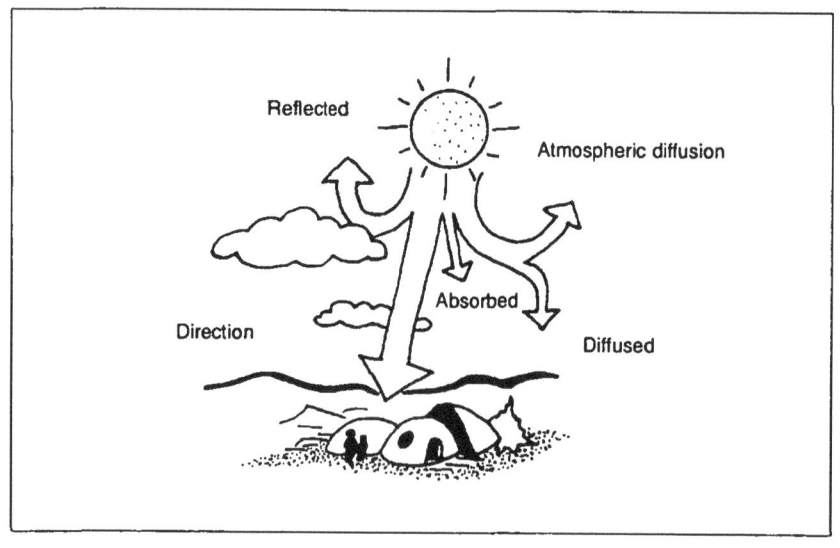

Figure 58 Qualitative distribution of solar radiation at the atmospheric layer

80

The impact of solar radiation on a given area

Several factors influence the intensity of solar radiation on a given area of the earth:

Composition of the atmosphere As we saw above, the presence of suspended particles and clouds causes reflection into space and diffusion of radiation. The composition of the atmosphere thus has an important part to play in the quantity and quality of energy received (and the proportions of direct and diffused radiation).

Thickness of atmosphere to be penetrated The thinner the atmospheric layer, the greater the radiation which reaches the surface. This thickness varies according to the altitude of the location; thus in mountainous areas the solar radiation is clearly greater. It also varies according to the time and day of the year, because of the annual oscillation of the earth and its daily rotation.

The angle of incidence between the surface and the direct radiation This angle determines the energy density received by the surface, with direct radiation defined as a series of parallel beams.

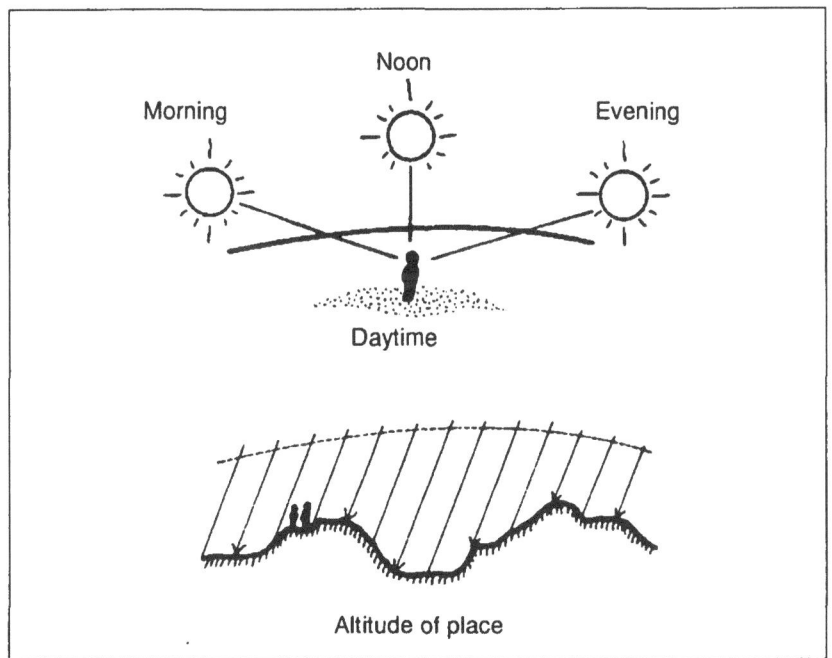

Figure 59 Variation factors affecting thickness of atmosphere to be penetrated

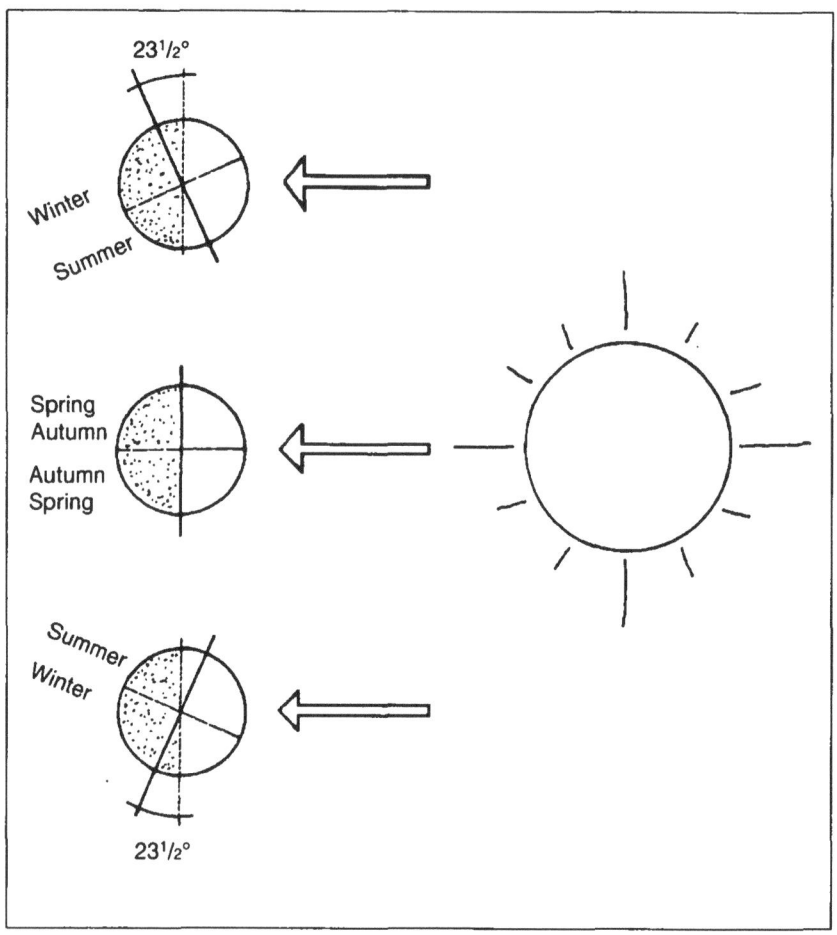

Figure 60 Seasonal variation factors *(according to E. Mazria)*

The angle of incidence is a function of several parameters:

o latitude of location;
o angle of inclination of surface to the horizontal;
o orientation of surface;
o time of day and day of year.

The albedo value of surrounding relief This reflective power concerns the overall radiation striking the relief of the area concerned. Part of this radiation, which is directly proportional to the surrounding albedo, is reflected back to the surface (see Table 1).

82

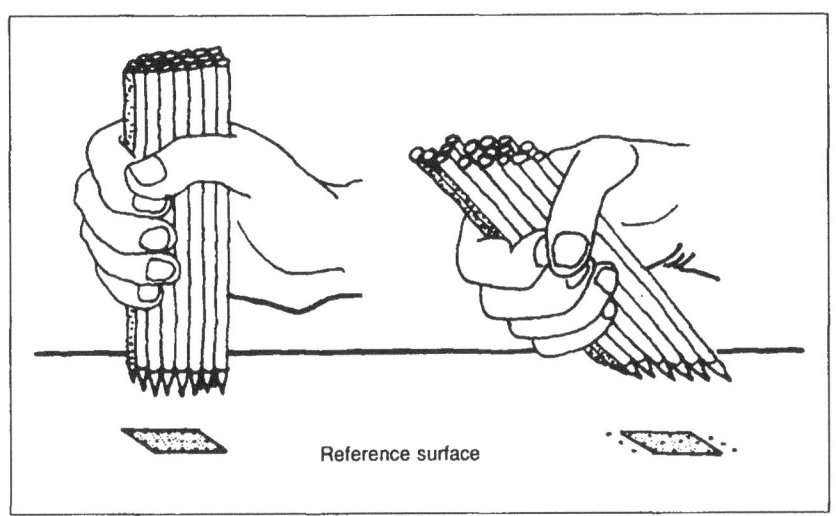

Reference surface

Figure 61 Demonstration of attenuation of intensity of overall energy received according to the angle of incidence – density of points of impact on the same receiving area is lower for inclined pencils *(according to E. Mazria)*

Table 1: Albedo value for the most common surfaces

Type of surface	Average albedo
New grass, lawn	0.20
Dry grass	0.30
Cultivated bare soil	0.16
Sandy soil	0.20
Clean sand	0.32
Gravel	0.22
Asphalt	0.19
Flat water (h > 30°)	0.04
Flat water (h < 10°)	0.45
Sea and ocean (h > 30°)	0.04
Sea and ocean (h < 10°)	0.10
Ice	0.30
Fresh snow	0.85
Packed or old snow	0.60

Possible masking   Masking refers to any obstacle between the source of radiation and the surface on which it impinges. It can be mountainous relief which, depending on the position of the sun in the sky, masks the direct radiation for part of the day, or it could be a nearby building.

Area of the receiving surface   The overall intensity of the energy received is directly proportional to the area of the surface receiving it.

The following table gives an overview of the effects of these different factors. This table gives a qualitative representation of available radiation intensity. What is interesting is that the chosen variables are independent (except for the albedo). Thus each criterion has a multiplying effect on the final value of the radiation received at the surface concerned. The presence of just one minus sign (–) causes a sharp fall in this value.

Table 2:   Factors affecting radiation

| Criteria | | Solar radiation received at surface | | |
		Strong	Medium	Weak
Atmospheric condition				
Clear sky	not polluted	+		
	polluted		0	
Cloudy sky	occasionally		0	
	consistently			–
Thickness of atmosphere				
Altitude	high	+		
	low		0	
Season	summer	+		
	winter			–
	autumn/spring		0	
Time of day	morning/evening			–
	midday	+		
Angle of incidence				
Orientation*	South	+		
	East/West		0	
	North			–
Degree	$0° < angle < 25°$	+		
	$25° < angle < 60°$		0	
	$60° < angle < 90°$			–
Albedo				
High (snow etc.)		+		
Medium (earth, grass etc.)			0	
Low				–
Obscurity (masking)				
None		+		
Morning/evening			0	
Middle of the day				–

* The recommendations given in the table refer to the northern hemisphere. For the southern hemisphere, it will be necessary to reverse the figures for the south and the north.

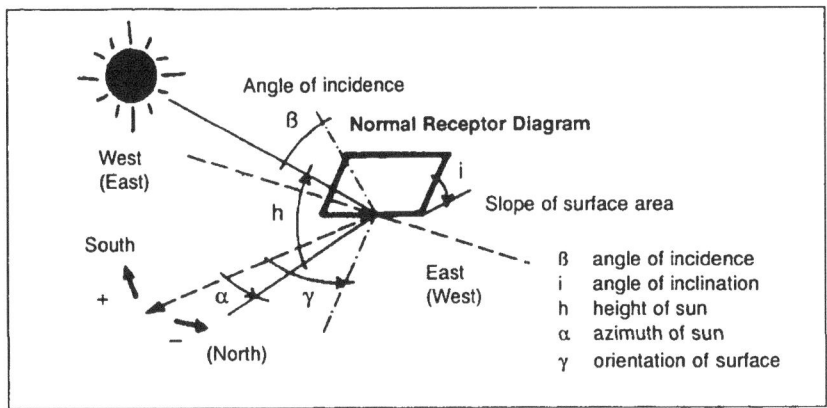

Figure 62 For the southern hemisphere, refer to cardinal points in brackets

Forecasting energy received at a given surface

For the designer it is important to anticipate what quantity of solar energy will be available for solar installations.

There are two possible ways of making these forecasts: either a meteorological survey is available which provides the necessary information in hourly or daily values of direct and diffuse radiation, or the quantity of solar energy must be calculated as a first estimate.

In the latter case a calculation method is needed. We present below a simplified method using graphical resolution, which allows rapid calculation of energy received at a given surface. It must be done in two stages.

Stage 1: Knowing the position of the sun at any time

For a particular place, the position of the sun is defined by its azimuth and its height.

These angles depend on the latitude, time and day of the year. The principle is to draw a diagram for a given latitude of the course of the sun in the sky for any day of the year.

To avoid having to accumulate 365 diagrams representing the changing position of the sun during a year, a single diagram shows representative positions for each month.

Appendix III gives a series of solar diagrams for latitudes between 0° and 50° north and south.

In this way the position of the sun at any moment of the year is now known with sufficient accuracy (see solar diagram for 30°N, Figure 64). We can now proceed to the second stage.

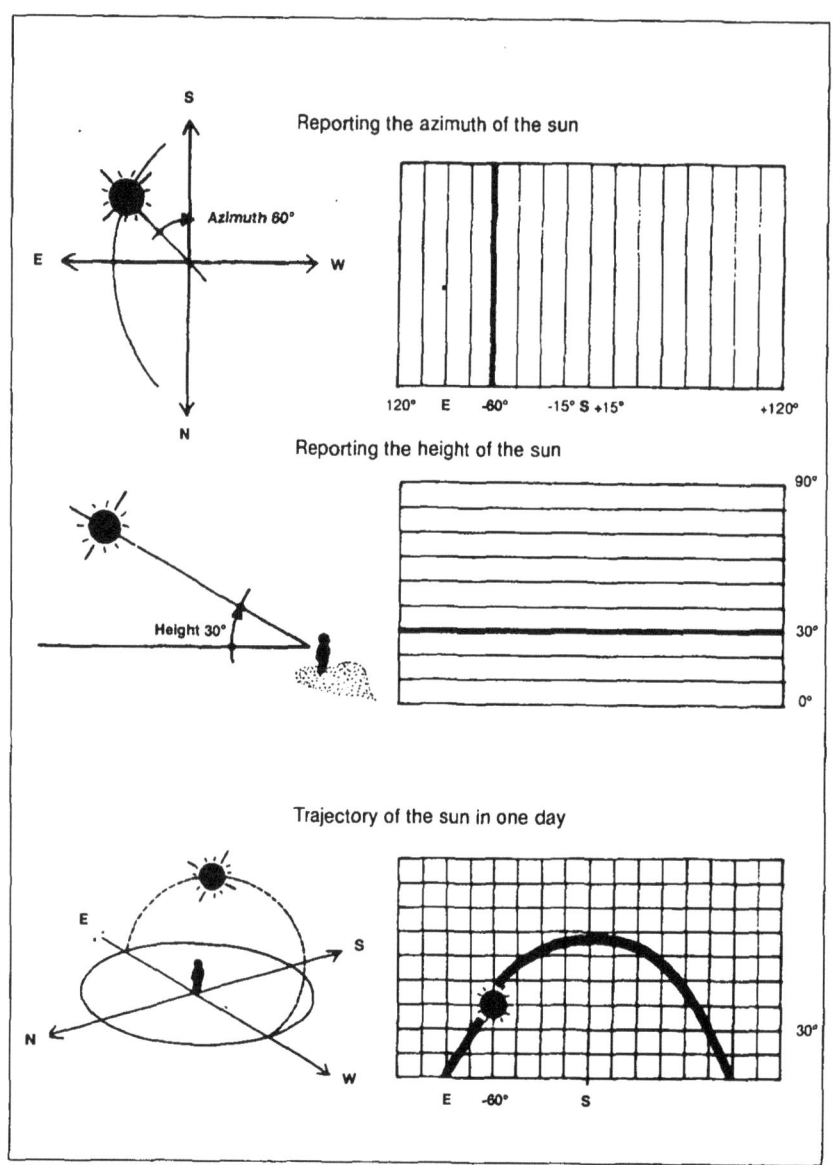

Figure 63 Drawing a solar diagram in rectangular co-ordinates

Stage 2: Calculating energy collected according to the position of the sun

Points of equal received energy intensity (power) per square metre (W/m^2) are drawn as a function of the height of the sun and the orientation of the sloping surface. These diagrams are called *radiation indicators*.

86

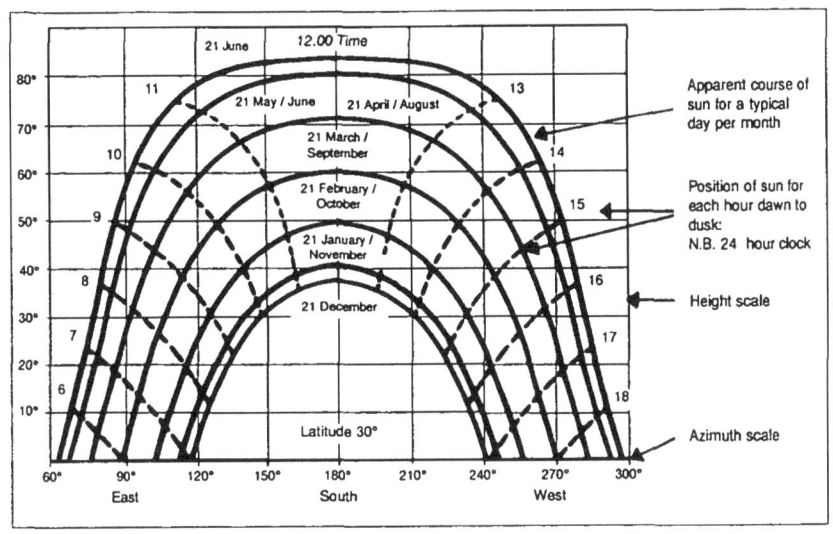

Figure 64 Monthly performance recorded over a year
(hours are given in true solar hours)

To calculate energy received as a function of the position of the sun, the indicator for the particular inclination merely has to be superimposed on the solar diagram for the latitude of the location in question. In this way, with a set of four indicators for surfaces of inclination which are: horizontal (= 0°), vertical ($i = 90°$), or sloping at 30° and 60°, it is possible to calculate hour by hour the energy received per m² of surface (in Wh/m²).

The various indicators are given in Appendix III.

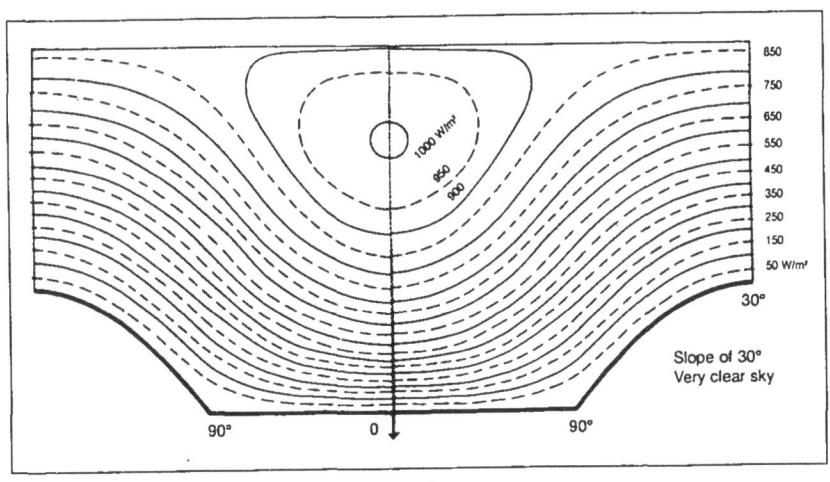

Figure 65 Radiation diagram for Leh

87

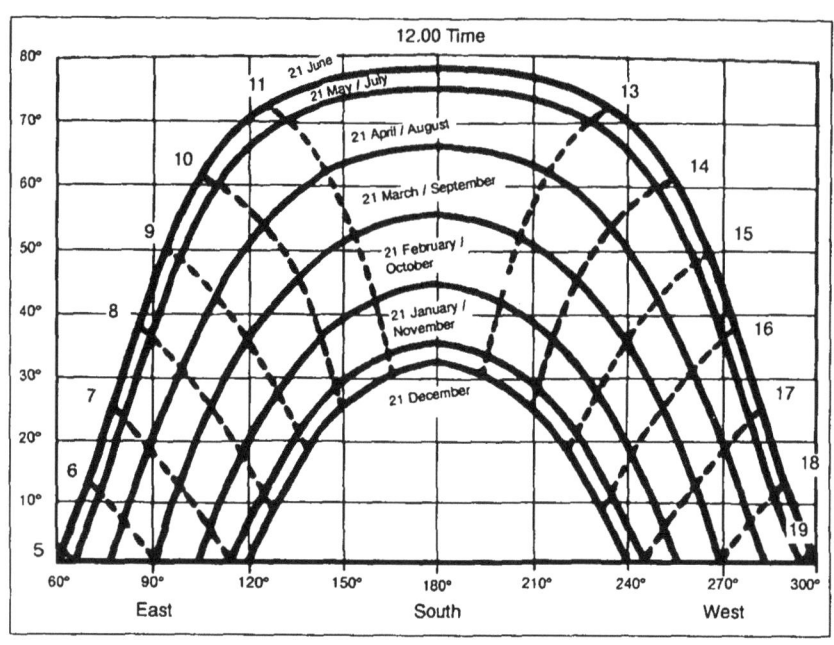

Figure 66 Solar diagram for Leh

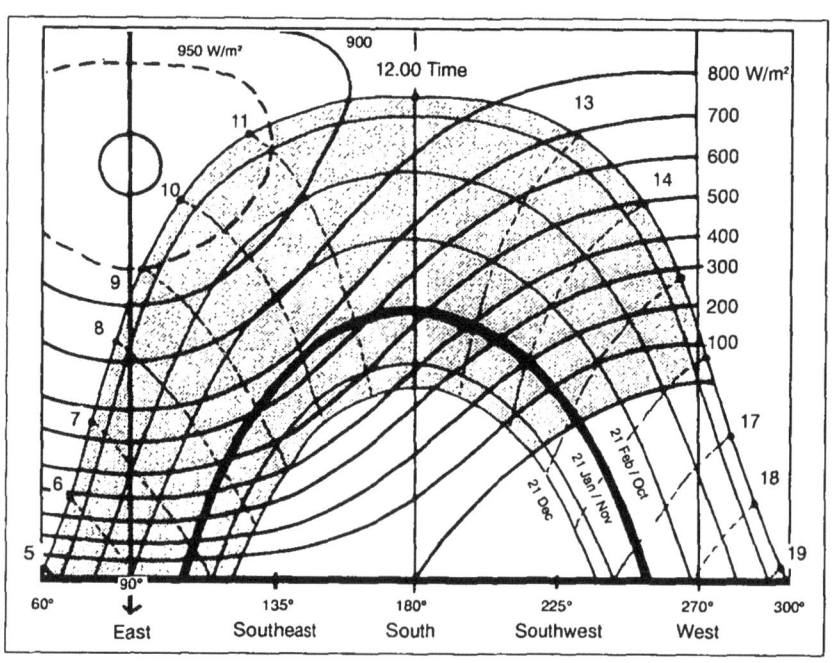

Figure 67 Indicator for Leh

88

Example of Leh (latitude 35°N)

The surface has a 30° slope, and the calculations are for a typical February day.

In the first instance the indicator and corresponding solar diagram are selected.

The indicator is then superimposed – previously traced onto tracing paper – on the cylindrical solar diagram, positioning the 0 of the indicator on the chosen orientation. In our example this is the 90° orientation of the solar diagram: due east.

Then proceed as follows to calculate the energy collected, hour by hour:

o *For a complete period of one hour*

The received power is read at the start of the hour (P1), in W/m^2. This is the intersection betwen the point which indicates the start of the hour on the solar diagram for the sun's progression on 21 February and the closest radiation curve.

The same procedure is followed for power at the end of the hourly period (P2) and total energy collected during this period is calculated using:

$E = (P1 + P2)/2$ in Wh/m^2.

o *For a period of less than one hour (often the case at sunrise or sunset)*

The method is identical, the average power is calculated (P1 +P2)/2 and is multiplied by the corresponding fraction of an hour in order to obtain the energy received in Wh/m^2.

In this example we obtain the data shown in Table 3.

Table 3: Energy collected over a day at Leh

Hours	Strength P1 (start)	Strength P2 (end)	Energy per hour (Wh/m²)
6.00– 7.00	0	150	(0 + 150)/2 × ½ hour = 37
7.00– 8.00	150	500	(150 + 200)/2 × 1 hour = 325
8.00– 9.00	500	600	(500 + 600)/2 = 550
9.00–10.00	600	650	(600 + 650)/2 = 625
10.00–11.00	650	650	(650 + 650)/2 = 650
11.00–12.00	650	550	(650 + 550)/2 = 600
12.00– 1.00	550	400	(550 + 400)/2 = 475
1.00– 2.00	400	200	(400 + 200)/2 = 300
2.00– 3.00	200	50	(200 + 50)/2 × 1 hour = 125
3.00– 4.00	50	0	(50 + 0)/2 × 1/5 hour = 5

Total energy received throughout a day in February = 3962 Wh/m²

6 Materials and their thermal properties

Now that it is possible to determine the sun's position and the energy received from it, the methods of absorption, storage and distribution of this heat can be learned. Firstly, however, the different modes of heat transfer must be understood.

Three fundamental modes of heat transfer

The heat or thermal energy of a natural medium is linked to the state of 'excitation' of its fundamental components: molecules, atoms, free electrons etc., which have a certain freedom of motion.

These elements are able to exchange energy, thus constituted, in different ways:

o by direct interaction with neighbouring particles (collisions): this is *conduction*;

o by mixing different parts of a fluid with different temperatures: this is *convection*;

o by absorption or emission of electromagnetic radiation: this is *radiation*.

In reality these three modes are interrelated.

1. Conduction

The molecules of the hottest part forcibly collide with each other and transfer this vibration energy (kinetic energy) to neighbouring molecules. Heat always flows from hot to cold parts.

The speed of propagation of the heat flow by conduction is the principal characteristic of this mode of transfer, called the *thermal conductivity* of a material, designated in this book by the Greek letter λ (lambda). It describes the material's capacity for receiving and transmitting heat.

2. Convection

Convection describes all heat energy transfer between a surface and a moving fluid in contact with it, or within a fluid with the movement of its molecules from one point to another.

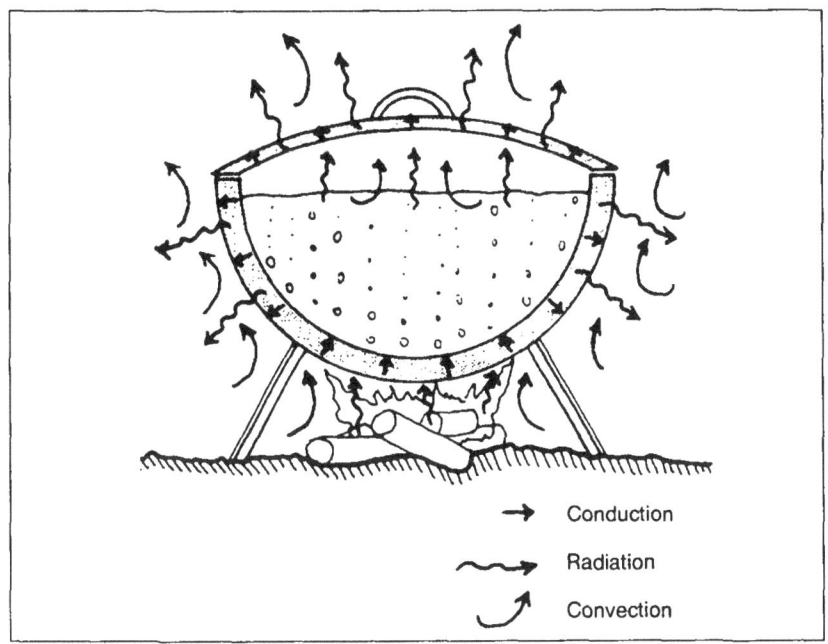

→	Conduction
～→	Radiation
⤴	Convection

Figure 68 Cooking pot on a wood fire

There are two distinct types of convection:

Natural convection   When molecules of a cold fluid such as water or air come into contact with a warm partition, part of the vibration energy causing movement in the surface molecules of the solid is transferred to those molecules of the fluid which are in close proximity. The quantity of heat transferred in this way is proportional to the temperature difference between the fluid and the surface of the solid.

Movement is created within the fluid: the fluid heats up, expands, becomes less dense and rises. It is replaced by cooler fluid and a rising movement is created. In contact with a cold surface, a warm fluid cools down and becomes heavier; this time a downward motion is brought about.

Forced convection   Mechanical means, such as a pump or fan, are used to promote rapid contact between the fluid and the partition. By rapidly replacing the mass of cold air on a warm surface, maximum temperature difference between the fluid and the partition is achieved. Consequently greater heat transfer is maintained than with natural convection.

In summary, to denote heat transfer by convection a *coefficient of surface heat exchange* (*h*) can be defined, which depends on several parameters:

- o characteristics of the fluid
- o flow rate
- o temperature difference
- o geometric arrangement

3. Radiation

Just like the sun, materials heated to a certain temperature emit energy to their environment in the form of electromagnetic waves (without material aid). The intensity of the radiation depends on the temperature of the radiating surface and its *emissivity*, ε *(epsilon)*.

A body radiating with an emissivity equal to one is called a black body. As an ideal emitter, it transmits maximum energy for a given temperature. In general, construction materials have a good emissivity: 0.9, that is 90 per cent of theoretical maximum. A coefficient of *absorption* α *(alpha)* is also defined to describe the proportion of the incident radiation which is absorbed by the material.

The steady state

From a thermal point of view every material is constantly seeking equilibrium between the heat absorbed on part or over all of its surface and the heat losses at its periphery. For each level of thermal equilibrium there is a corresponding level of heat and temperature of equilibrium (see Figure 69, the leaking bucket analogy).

The thermal behaviour of materials

Several categories of construction materials have been described according to their thermal properties. These include opaque, insulating, transparent and translucent materials and materials which change phase.

Opaque materials

These are materials which only allow transfer by conduction. In our research they involve dividing walls in houses, which do not transmit solar radiation directly to the inside. In the study of heating buildings, a *coefficient of thermal transmission of a partition (K)* has been identified. This involves heat flow transmitted across a square metre of dividing wall per one degree temperature difference between the two environments separated by this wall. This coefficient is thus expressed in watts per square metre per degree ($W/m^2/°C$).

To calculate the coefficient, the partition is broken down into its different materials. For each layer identified in this way, there is an associated thermal resistance (R). The value of R is equal to the thickness of this layer divided by its thermal conductivity.

For films of air between ambient temperatures and surface temperature a resistance for this film of air can be:

$$R = \frac{1}{h_{ext} \; or \; h_{int}}$$

'h_{ext}' or 'h_{int}' is the coefficient of convection for the exterior (h_{ext}) or interior (h_{int}) surfaces of the building. These values are given in Appendix VI.

For any partition:

$$R_{total} = R_{ext \; air \; layer} + R_1 + R_2 + ... + R_n + R_{int \; air \; layer} \; \text{with } R=e/\lambda$$

$$K \; overall = 1/R \; total \; (W/m^2 \; /^{\circ}C)$$

Warning: K overall is not equal to the sum of the Ks for each layer.

Thus, over time, only the overall thermal resistance of a partition is required to describe the heat flow between exterior and interior.

In fact, in a natural environment, exterior and interior conditions are constantly changing, for many reasons: rise and progression of the sun, variation in outside temperature, interior heating etc.

Another important parameter characterizes thermal inertia of a material: *thermal mass*, also called specific heat (symbol Cp), and defined as the quantity of heat needed to raise the temperature of the material by one

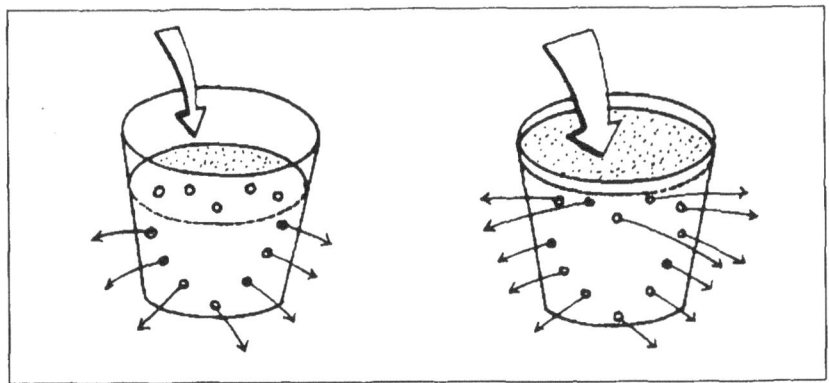

Figure 69 Analogy of the leaking bucket *(greater heat level gives rise to increased temperature and greater thermal losses)*

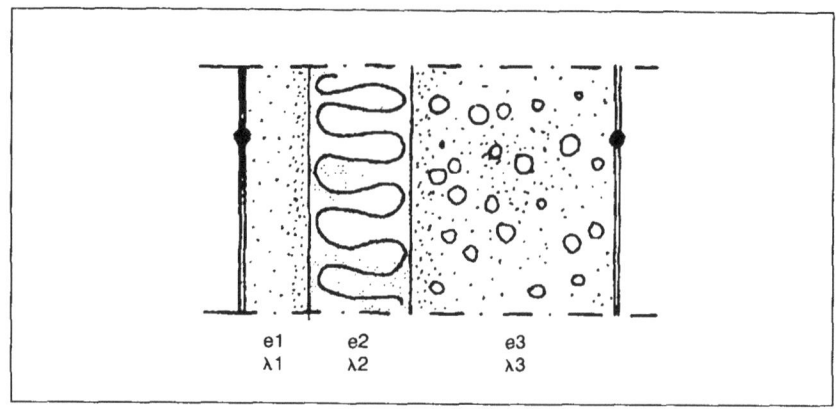

Figure 70 Calculating coefficient K for a partition (steady state) *For each layer a thermal resistance is defined (R = e/λ) (e = thickness of the layer in question and λ its conductivity). Then the resistances of the external and internal air layers are added together. To obtain the overall K all that is needed is for the inverse of this sum to be calculated.*

degree, expressed in joules per kilogram per degree (J/kg/°C), or in watt-hours per kilogram per degree (Wh/kg/°C).

Thus knowledge of the three parameters λ, Cp and the *density* (symbol ρ) is sufficient to calculate thermal behaviour of an opaque partition.

> with damping $= \dfrac{\theta_x}{\theta_o} = \exp\left[-x.\sqrt{\dfrac{\pi}{a.24\ hours}}\ \right]$ and time lag $= \dfrac{1}{2}.x.\sqrt{\dfrac{24\ hours}{\pi.a}}$
>
> $a = \lambda.\rho.$
>
> Cp is called the admissivity of the material

With an exterior surface temperature which varies according to a sinusoidal law of a 24-hour period (the case for a sunny day), the temperature of the interior area of the wall in question varies according to a sinusoidal law for the same period, but it has been subject to damping in amplitude and a phase lag, both of which are directly linked to its thermal characteristics (λ, ρ, Cp) for a given thickness.

Insulating materials

For opaque partitions insulating materials are defined as having the characteristics K < 0.12 W/m²/°C. These materials play a decisive part in the efficiency of passive solar systems by limiting thermal losses at each stage

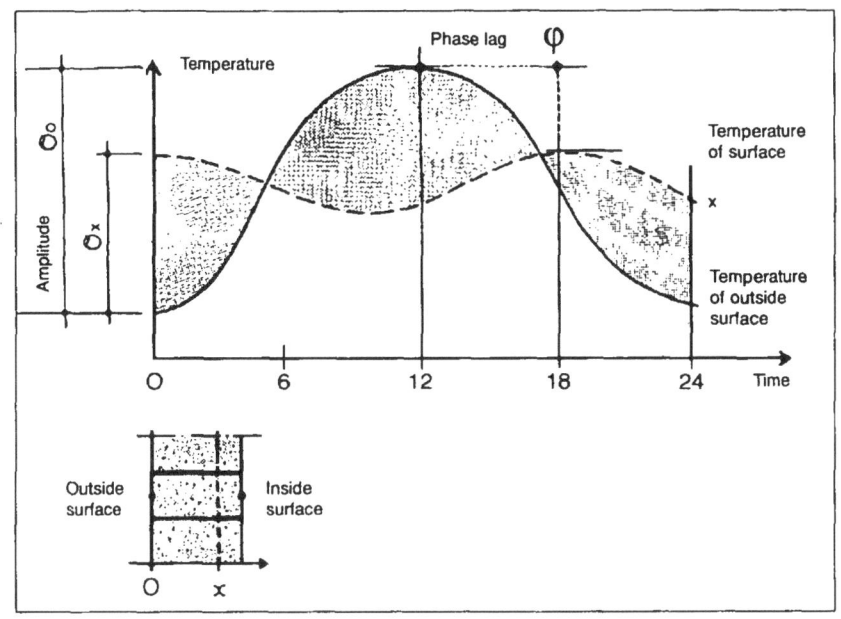

Figure 71 Concept of time lag in systems in dynamic mode

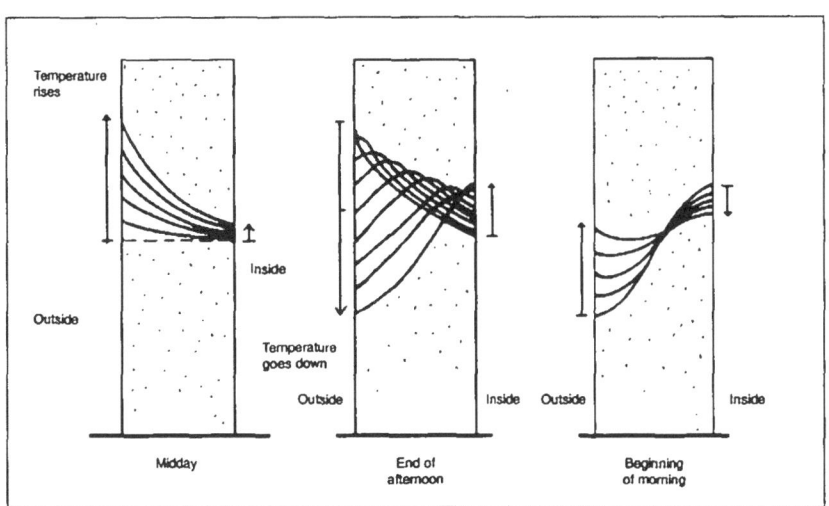

Figure 72 Phase lag of temperature levels through an opaque wall
from Passive Solar Design Handbook, *Washington, 1980*

95

Table 4: Characteristics of different insulants

Name	Thermal conductivity (W/m/°C)	Fire resistance (without coating)	Resistance to temperatures up to 100°C	Resistance to termites	Resistance to rodents	Resistance to humidity	Cost
Groundnut shells or woodshavings	≈ 0.10	−	+	−	−	0	−
Compressed straw	≈ 0.10	−	+	−	−	−	−
Cork	0.04 ↔ 1.10	−	+	−	+	+	+
Glass wool or rockwool	0.03 ↔ 0.01	+	+	+	0	−	0
Lightweight, resinous wood	≈ 0.12	−	+	−	+	0	+
Plywood	0.12 ↔ 0.15	−	+	−	+	+	+
Vermiculite	≈ 0.05	+	+	+	+	+	+
Expanded polystyrene	≈ 0.04	−	−	−	−	+	+
Polyurethane, blown polystyrene	0.04 ↔ 1.10	−	+	+	0	+	+

+ = high   0 = medium   − = low

from R. Celaire

(collection, storage or distribution). They thus ensure maximum use of heat absorbed.

Transparent materials

These are materials which transmit solar radiation. They are described by three parameters: t = level of transmission of incident radiation; r (ρ) = level of reflection and a (alpha) = coefficient of absorption.

Table 5 shows a variety of transparent materials.

For a certain quantity of incident solar radiation – direct and diffuse – part is reflected depending on the angle of incidence, part is absorbed in the material, causing its temperature to rise, and the rest is transmitted to the interior. Added to this is the part which is re-emitted to the outside and inside by the material when it warms up (thermal radiation).

The solar factor is defined as the ratio of transferred to incident energy.

The angle of incidence (between the normal to the glass and the radiation) has considerable influence on the quantity of energy transferred, particularly over 50°.

The greenhouse effect

This characteristic of glass makes it a basic material for the majority of solar systems. In fact most solar radiation is transmitted through a pane of glass. This radiation heats the inside surfaces of the glazed area, their temperatures rise and long wavelength thermal radiation is directed at the glass. This glass is totally opaque to this range of wavelengths.

Phase change materials

Some materials demonstrate one interesting characteristic for normal temperatures in passive solar systems: they change their phase (solids becoming liquids, for example). In order to change from one phase to another, a considerable amount of energy called *latent heat* is needed. Such materials are therefore very useful for storing energy in a restricted volume.

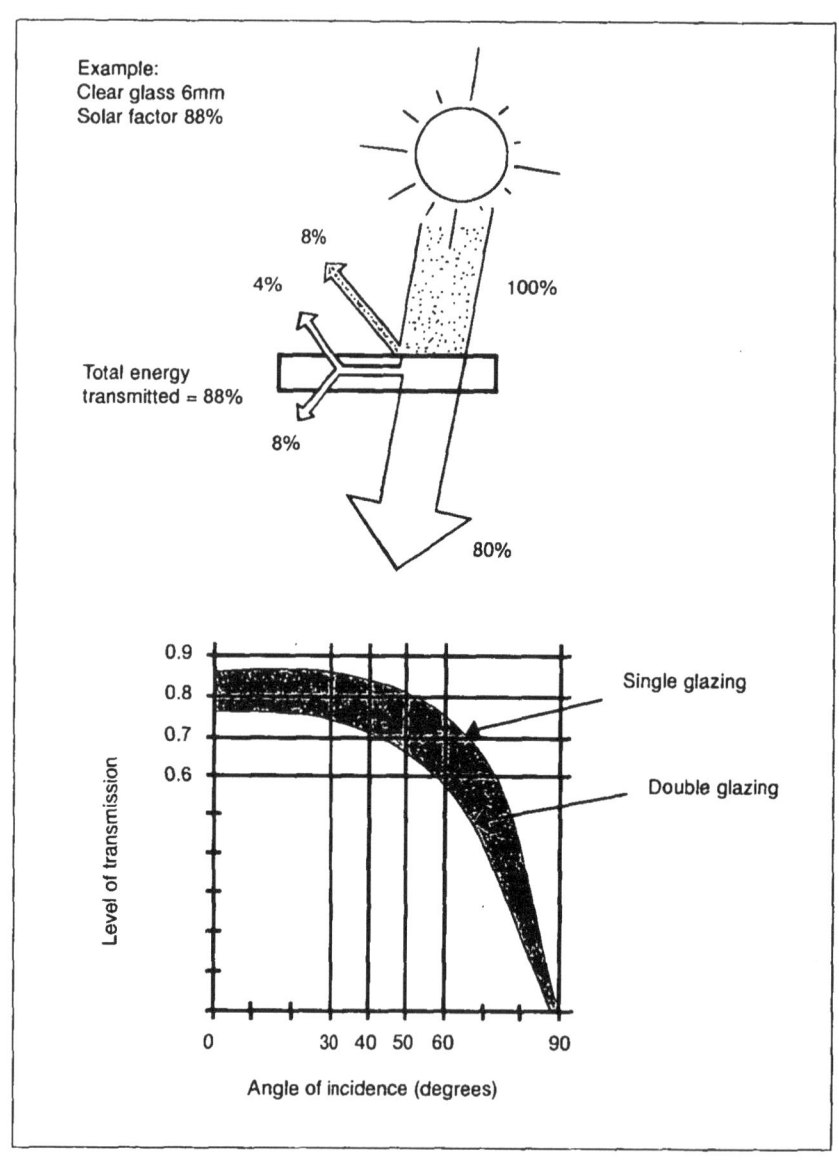

Figure 73 Effect of angle of incidence on heat transmitted

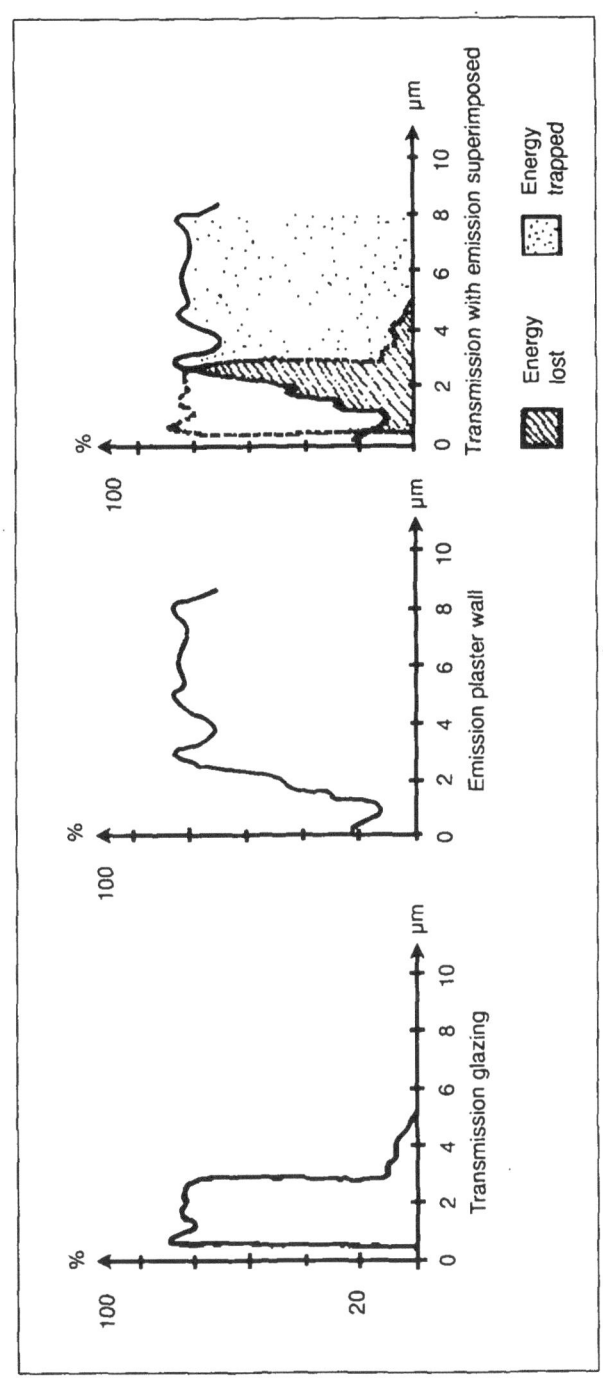

Figure 74 Trapping thermal radiation through glass

99

Table 5: Properties of different transparent coverings

Materials	Ordinary glass	Tempered glass	Strengthened glass	Polyester glass fibre composite	Polycarbonate double thickness	Polypropylene double thickness	Polymethacrylate double thickness (acrylic)	Polyethylene (1)
Solar radiation transmitted under normal conditions (%)	86/91 (4mm)	86 (4mm)	72	83	80	65	83	85 (75)
Thermal transmission (%)	3	3	3	6	6	approx. 10	6	70 (30)
Transparent (P) Translucent (L)	P	P	P	L	L (single thickness: P)	L	L (single thickness: P)	L
Standard thicknesses available	3 to 6mm	4 to 8mm	6mm	1mm	4 to 16mm	4mm	8–16mm	0.1–0.3mm
Maximum temperature used	200°C	200°C	200°C	70°C	100°C+	approx. 60°C	90°C	60°C
Fire resistance	good	good	good	poor	medium	poor	poor	very poor
Weight kg/m^2	(4mm thick) 11	(4mm) 16	(6mm) 16	(1mm) 1.5	(6mm) 1.2	(4mm) 0.7	(16mm) 5	(1mm) 0.17

(1) Values in brackets are for polyethylene treated for UV (from B. Yanda).

Table 5 (contd.)

Expansion	low	low	high	high	high	high	very high
Length of life (beyond which transmission of solar radiation is 90% of initial value)	100 years +	100 years +	7–10 years (or around 15 years with UV protection)	6 years	around 2 years (very poor)	15 years +	1 to 3 years
Usual fixing method	putty, battens	putty, battens	staples, nails	special profile or putty, battens	special profile or putty, battens	special profile or putty, battens	staples
Tools	diamond hammer, putty knife	diamond hammer, putty knife	staples, nails	saw	saw putty, battens	saw putty, battens	scissors, staple gun
Costs (2)	*	**	*	*	**	**	0

(2) Good value: 0; medium price: *; expensive: **

Table 6: Fusion temperature and latent heat of fusion for various materials

Material	Fusion temperature °C	Latent heat kJ/kg
Tabromide	32	37
Metaphosphoric acid	43	107
Beeswax	62	177
Potassium hydroxide	63	63
Phosphoric acid	70	156
Paraffin	74	230
Naphthalene	80	149
Aluminium bromide	97	42

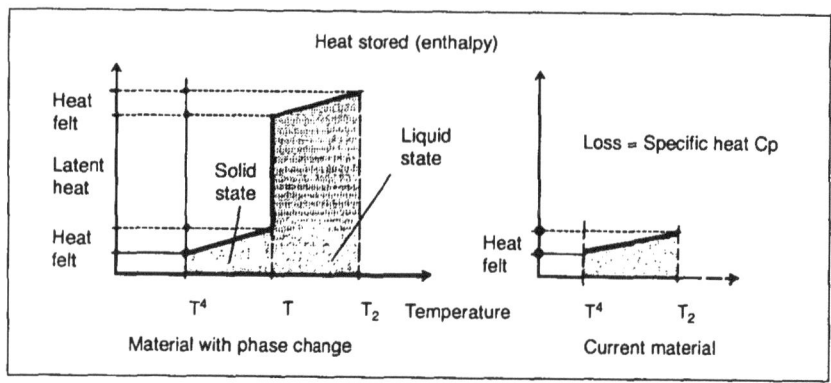

Figure 75 Characteristic curve of increase in stored heat as a function of temperature of the material

Table 7: Thermophysical properties of materials

Material	Thermal conductivity $(W/m^2/°C)$	Density (kg/dm^3)	Specific heat	Coefficient of absorption $(Wh/kg/°C)$	Coefficient of emission
Air (20°C)	0.024	$1.29-10^3$	0.28		
Aluminium	221	2.74	0.25	0.04	0.09
Ashes	0.071	0.64	0.23		
Asphalt	0.74	2.11	0.26	0.9	0.9
Bakelite	16.78	1.3	0.41	0.9	0.9
Brick (baked)	0.7	1.97	0.23	0.68 (red)	0.9
Cardboard	0.07				
Cement (Portland)	0.029	1.92	0.19	0.6	0.9
Charcoal	0.052	0.24	0.23	0.9	0.8
Cork	0.048	0.09	0.56		
Concrete (Standard)	1.2	2.31	0.18	0.6	0.9
Cotton (fibres)	0.042	1.52	0.37		
Earth (dry, compacted)	0.64-1	1.5-1.9	0.23	0.75	
Fibreglass	0.038	0.05	0.18		
Glass	1.028	2.47	0.2		0.84
Ice (0°C)	2.25	0.92	0.56		0.95
Iron (cast)	48	7.21	0.14	0.3-0.8	
Lime	0.93	1.65	0.25	light 0.35 dark 0.5	
Mild steel	45.3	7.83	0.14	0.3-0.8	0.12
Paper	0.13	0.93	0.37		0.9
Paraffin	0.24	0.9	0.80		
Rock	2.5	2.6	0.25	0.4-0.6	0.9
Sand	0.33	1.52	0.22	0.8 (dry)-0.91	
Sawdust	0.06	0.19		0.4	
Thatch	0.09	0.27	0.28		
Water (20°C)	0.6	1	1.16		
Wood	0.11-0.25	0.37-1.12	0.5-0.75	0.6	0.9
Wool fibres	0.04	1.31	0.38		

AIR

Temperature °C	Specific heat at constant pressure J/kg/°C	Conductivity W/m/°C
0°	1 004	17.1910^{-3}
20°	1 006	19.2610^{-3}
40°	1 010	21.2410^{-3}
60°	1 025	25.1210^{-3}

Table 7 (contd.)

80°		1 045	28.8610^{-3}
100°		1 069	32.4510^{-3}
200°		1 092	35.7010^{-3}
300°		1 184	49.3310^{-3}

WATER

Temperature	Density	Specific heat at constant pressure	Thermal conductivity
°C	kg/m^3	J/kg/°C	W/m/°C
°0	1 000	4 220	0.55
20°	998	4 183	0.60
40°	992	4 178	0.63
60°	983	4 191	0.65
80°	972	4 199	0.67
100°	958	4 216	0.68
200°	863	4 501	0.67
300°	700	5 694	0.56

COEFFICIENTS OF SOLAR ABSORPTION

An approximate value can be estimated by the colour of the surface (flat, smooth colour)

Colour	Coefficient
White	0.25 to 0.4
Grey to dark grey	0.4 to 0.5
Green, red, brown	0.5 to 0.7
Brown to dark blue	0.7 to 0.8
Dark blue to black	0.8 to 0.9

COEFFICIENTS OF SURFACE EXCHANGE ON VERTICAL WALLS

Speed of wind (m/s)	Coefficient of surface exchange	
	W/m^2/°C	kcal/m^2/h/°C
0 (natural convection)	8.2	7.1
2.0	18.1	15.6
3.5	22.7	19.5
4.5	26.1	22.5
6.5	34.0	29.3
9.0	41.4	35.6
11.0	48.8	43.0
13.5	56.8	48.8

Source: Clifford Strock and Richard L. Koral. *Handbook of air conditioning, heating and ventilating*

7 Details of collection, storage and distribution

Every building creates an internal micro-climate providing a certain level of comfort for the occupant. When passive solar systems are used, there is a direct relationship between exterior and interior climate.

The *bioclimatic* design (see Figure 76) of these systems enables an acceptable level of interior comfort to be maintained based on external conditions (sun, temperature, wind, humidity, etc.).

1. Picking up
2. Storing
3. Insulating
4. Distributing

Figure 76 Bioclimatic design

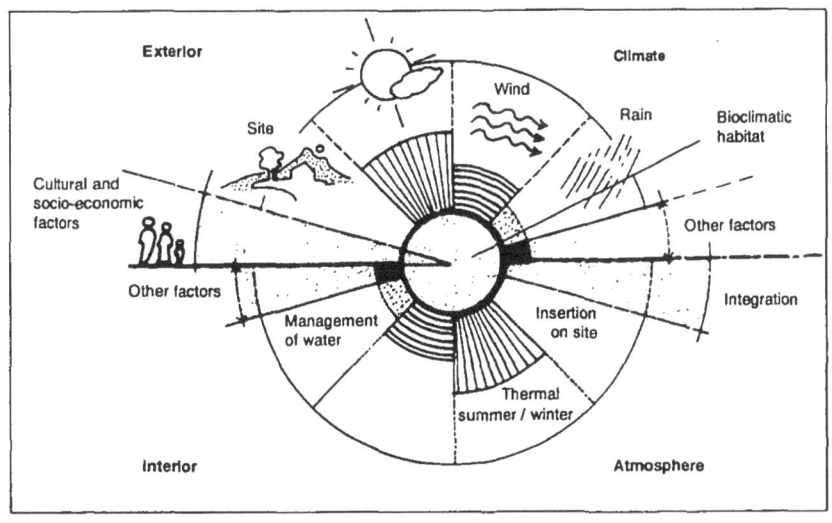

Figure 77 Bioclimatic design of a dwelling

For heating during the winter a balance should always be struck between *absorption* of solar radiation, *storage* of received heat and *internal distribution* of stored heat, taking particular care with thermal insulation at each stage.

There are two types of solar equipment:

o *Those which provide indirect solar heat gain (the concept of sun/space/ mass)*
Solar energy enters the building directly through a large glazed area (bay window, greenhouse) and heat is stored in the sides of the building (walls, ceiling or floor). This method combines a large glazed area facing the equator (south-facing for the northern hemisphere) connected to adequate thermal mass on the inside.

o *Those which give rise to direct heat gains (concept of sun /mass/space)*
Solar energy reaches thermal mass situated between the sun and the building to be heated, such as the greenhouse-effect wall, roof tank, adjoining greenhouse etc.) The thermal mass, called the collector-accumulator wall, absorbs solar radiation, transforms it into heat and releases it with a particular phase lag and damping.

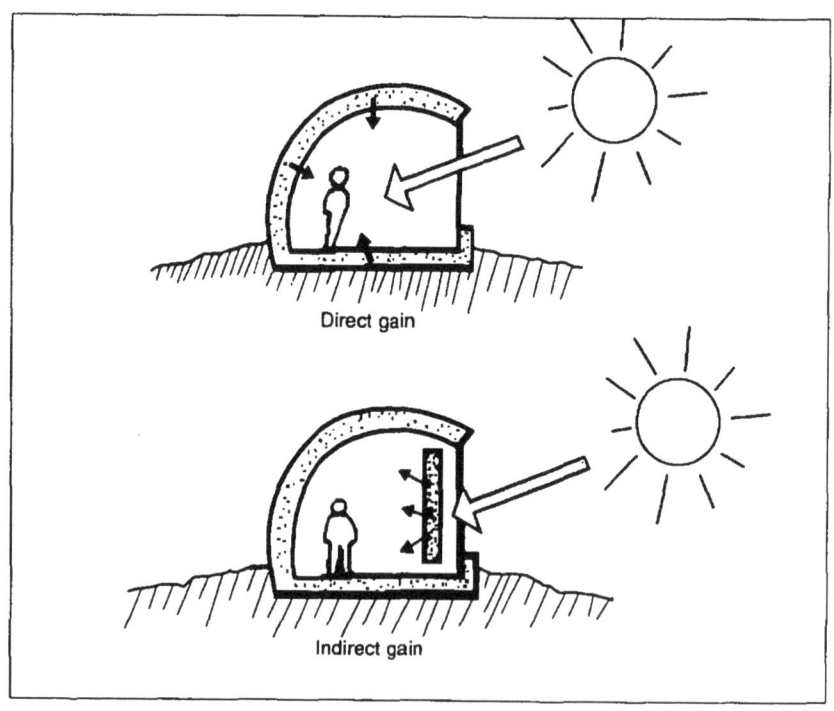

Direct gain

Indirect gain

Figure 78 Direct and indirect gain

106

Absorption

For absorption, glazing and plastic transparent sheets are essential components.

As energy transfer is influenced by the angle of incidence between the solar radiation and the glazed parts, the following rules are applied for the angle of inclination of the collector area when winter heating is applied:

Latitude of site	Angle of inclination
0° to 10°	$i = 10°$
10° to 20°	$i = $ latitude
20° to 35°	$i = $ latitude $ + 10°$
more than 35°	$i = $ latitude $ + 15°$

The aspect should always be due south if possible for the northern hemisphere and due north in the southern hemisphere.

Of course it is important to avoid anything masking the collector areas too much. The following recommendations are made:

Latitude of site	North	South
0° to 10°	$b < 67°$	$a < 55°$
10° to ± 20°	$b < 77°$	$a < 45°$
20° to ± 30°		$a < 35°$
30° to ± 40°		$a < 27°$
40° to ± 45°		$a < 17°$
Obstacles	East/West $c < 25°$	

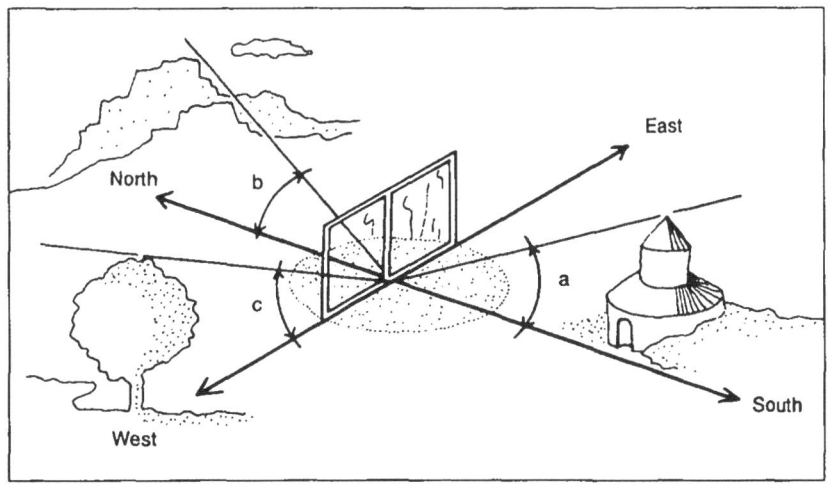

Figure 79 Distance from obstacles

107

The type of frame (holding the transparent parts) plays an important part in the overall coefficient of thermal transmission, K. A metal frame for example is a greater source of heat loss than a wooden frame, and causes noticeable reduction in performance of the system.

Table 8: Different values of K according to the type of glazing (W/m² °C)

Type of glazing	Thickness of layer of air (mm)	Without frame	Wood frame	PVC frame	Frame without thermal layer	Aluminium frame with thermal layer
Single	–	5.5/6.7	4.5/5.3	4.3/5.1	5.8/6.6	5/5.8
Double	3	4.0	3.3	3.3	4.7	3.8
	6	3.4	2.9	2.9	4.3	3.4
	12	3.0	2.6	2.6	3.9	3.1
	≥20	2.9	2.5	2.6	3.9	3.3
Triple	3	3.0	2.6	2.7	4	3.1
	6	2.5	2.3	2.3	3.6	2.7
	12	2.1	2	2.1	3.3	2.5
	≥20	2.0	1.9	2	3.3	2.4
Thermal resistance frame alone		–	1.7	2	6.5	3.4
Percentage of surface glazed		100	72.5	65.6	71.8	71.8

Storage

Storage takes place as sensible heat. It is mainly located in the heavy masonry parts or containers of water, such as 200-litre barrels for example. The materials used for absorption must have high thermal capacity (maximum for water with $C_p = 4.19$ kJ/°C).

Trials under real conditions (Los Alamos Scientific Laboratory, New Mexico) proved the importance of direct exposure in storage mass. A thermal mass directly exposed to the sun is three to four times more effective for storing heat than the same mass placed in the shade within the same building. It is also important to ensure that the area exposed to the sun is dark in colour, to achieve maximum absorption of incident radiation.

Apart from specific thermal characteristics of storage materials, the size and shape of the store play an important part in releasing heat.

For example, 200 one-litre containers will not have the same effect as a single barrel of 200 litres. Small containers absorb more energy per unit of

Table 9: Some examples of typical storage materials

Material	Specific heat kJ/kg/°C	Density kg/dm^3	Thermal capacity kJ/dm^3/°C
Adobe brick	0.92	1.44	1.34
Brick	0.84	1.92	1.61
Concrete	0.96	2.40	2.31
Earth	0.88	1.52	1.34
Sand	0.84	1.76	1.48
Steel	0.50	7.85	3.96
Stone	0.88	2.64	2.32
Water	4.19	1.00	4.19
Wood	1.38	0.51	0.71

Source: B. Yanda, *Une serre solaire pour chauffer votre maison et pour jardiner toute l'année*, French translation, Edition Eyrolles, 1982

volume because they have a greater area to volume ratio. On the other hand they lose heat more rapidly than they accumulate it.

A system of long-term storage requires two to three days' sunshine to store sufficient heat, while short-term storage only requires one day. A system combining the two can deal with a cloudy period of two to three days, or short spells of sunshine.

Heat distribution

Releasing heat stored at night, or redistributing heat which comes in during the day, determines effective performance of passive solar heating systems. Distribution of this heat occurs in two ways: *convection* and *radiation*.

In direct heat gain systems, short-wavelength solar radiation which enters the building through the glazing undergoes a whole series of changes. Some is absorbed, some converted into heat and some reflected according to its angle of incidence and the colour of the interior surfaces.

If the whole interior surface is light in colour, there will be efficient distribution of heat. If this surface is dark, there will be a high concentration of this heat in the area where radiation is first incident.

Once the short-wave radiation is transformed into heat after its first incidence, three things can happen:

o heat travels through the material;
o heat is retransmitted in the form of infra-red to the inside of the building;
o heat is transferred by means of convection into the air of the building.

Most of the energy is transferred by infra-red radiation. Each surface not directly exposed to the sun's rays is continuously exposed to infra-red rays originating from the radiating surface according to its *angle of vision* with the latter.

On the other hand, it is preferable to maintain energy in the form of infra-red radiation or to store it in dense materials. Inevitably part of the energy is transferred to the air which heats up much more rapidly than thermal masses.

Convection currents are then created, accounting for about a third of the energy entering the building. They allow the heat to reach parts not affected by infra-red radiation.

In the case of convective heat transfer, heat exchange between two spaces should be possible to calculate in two ways:

o by a simple opening

$$Q_{Exchanged} = 44.L.(H.\Delta T)^{3/2} \text{ in Watts}$$

$Q_{Exchanged}$: The quantity of heat transferred between two spaces
L: Width of door (in m)
H: Height of door (in m)
ΔT: Temperature difference between rooms (in °C)

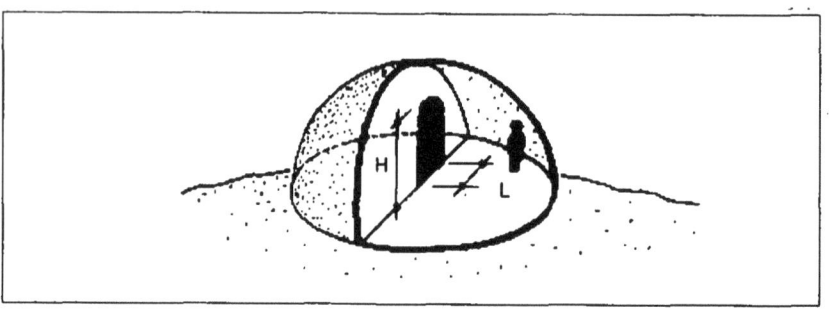

Figure 80 Simple opening

o for two openings in the wall

$$Q_{Exchanged} = 154.A.H^{0.5}\Delta T^{1.5} \text{ in Watts}$$

$Q_{Exchanged}$: quantity of heat exchanged between two rooms
H: height of the two openings (in m)
A: total area of the openings (in m^2)
ΔT: Temperature difference between two rooms (in °C)

110

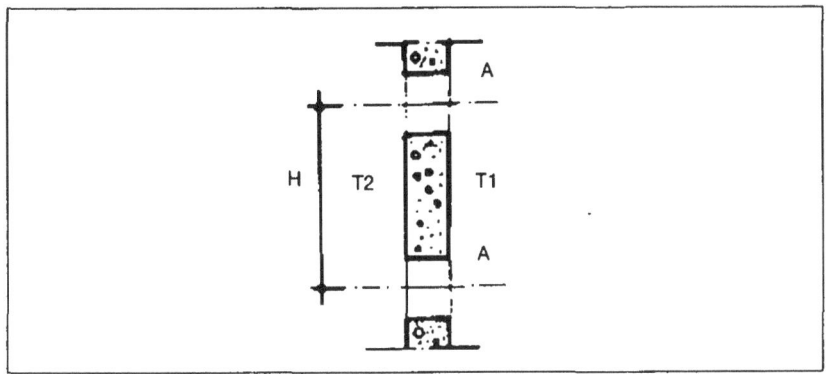

Figure 81 Two openings

Collector-storage interaction

Every passive solar system is a balanced combination between these three
factors: collection, storage and distribution. *Interior environment* thus
determines all parameters governing the level of comfort achieved.

In the first approach, two basic parameters are considered: the intended
inside temperature and the level of daily variation of the inside temperature. The American standard defines a level of discomfort as amplitude
greater than 6°C.

The role of inertia

In passive systems two criteria, desired temperature and variation in
interior temperature, often conflict. In fact, in order to achieve a certain
level of temperature rapidly, the collector area will be given priority. If the
inertia, that is to say the storage capacity of the system, is insufficient, the
temperature rises in proportion to the incident radiation. At night the
reduced storage will not be sufficient to maintain an acceptable temperature.

On the other hand, if the inertia of the building is increased, the
maximum temperature will be lower, but when the sun is not shining (at
night or during cloudy periods), the minimum temperature will be mark-
edly higher.

Thus, for a single average temperature over 24 hours, a building will be
more or less comfortable, depending on its inertia.

Interior inertia allows variation in interior air temperature of a building
to be reduced. It can be used to delay the contribution of daytime heating
and release heat in the evening, in a day room, a kitchen or a bedroom, for
example. This is the principle of the collector/storage walls, whose thickness is calculated according to intended use.

111

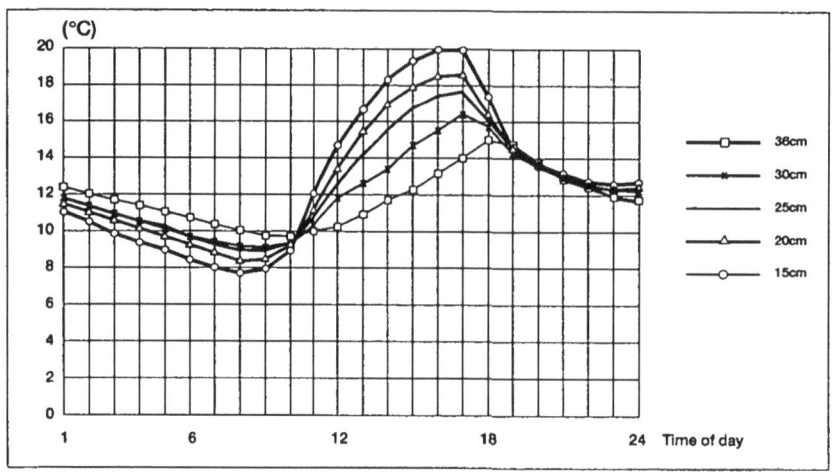

Figure 82 Influence of solar wall thickness on inside temperature

There is a limit beyond which added inertia has no further effect on inside temperature. Apart from this quantitative aspect it is important to emphasize the positioning of inertia. For direct heat gain, systems which distribute inertia equally around the building prove to be more effective than if inertia is concentrated on one wall. A greater heat exchange area provides greater storage/de-stocking capacity.

8 Role of thermal insulation

Intuitively one would think that to increase the inside temperature all that is needed is sufficient increase in collector area, and hence interior heat gain, according to storage mechanism. This condition is, however, found to be insufficient, and another basic parameter comes into play, particularly at night or during cloudy periods: *insulation*.

Insulation helps to limit losses, either by conduction through opaque walls, or by infiltration/exfiltration of air between the inside and outside of the building. There are several measures that limit losses from a building:

o thermal insulation of glazing
o controlling air recirculation
o insulation of opaque partitions (walls and ceiling).

Thermal insulation of glazing

This is used for the most conductive surfaces of the house, particularly the glazed areas (K of a glass panel = 5 $W/m^2/°C$ compared with K = 1$W/m^2/°C$ for the masonry parts, which are not insulated).

In fact any increase in total glazed area should be avoided, because beyond a certain limit the resulting losses – especially at night – exceed energy gained by this increase during the day. This implies that over 24 hours the heat load will be lower and hence average inside temperature will be lower (see analogy of the leaking bucket).

To insulate bay windows, two standard insulation methods can be used: double glazing, which consists of adding a second glass pane, separated from the first by a layer of air; or placing an insulating cover (shutters etc.) between the outside and the single glazing.

The second solution gives excellent results and is still the simplest. Nevertheless it does require daily maintenance by the user, and it is not an obvious solution for a glazed part which is in front of a collector/storage wall. In fact this area remains hidden from the interior of the building and it seems less natural for the user to place an exterior insulating cover in front of a collector/storage wall than to close shutters over a window at sunset.

From experience, therefore, using double glazing is preferred for greenhouse walls and less accessible glazed parts such as skylights.

Controlling air recirculation

Once glazing is insulated it is important to control exchange of air with the outside.

Air exchange is measured by hourly volume of new air. In the regions studied, air exchange leads to significant thermal losses, because the outside air is very much cooler than the inside air.

Part of the heat in the building is stored in the mass of cold air which enters (infiltration). This is then lost when this mass of air is released to the exterior (exfiltration). These movements are expressed as a function of the number of times the total volume of air in the building is replaced every hour, that is:

$$\text{Losses} = 0.34.N.V.\ \Delta T$$

ΔT: temperature difference between inside and outside (°C)

0.34:   ($Wh/m^3/°C$) thermal capacity of a m^3 of air

N: hourly rate of air replacement

V: volume of inside air in m^3

Table 10: Hourly level of air replacement in homes

Room or building	Hourly rate N by volume/hour
Room without door or window	0.5
Room with external door or window on one side	1
Room with external doors or windows on two sides	1.5
Room with external doors or windows on three sides	2
Entry hall	2

Source: ASHRAE – *Handbook of Fundamentals*, 1977.

Take two-thirds of these values for buildings with draught-proofing, or particularly airtight windows.

User behaviour has a great effect on losses caused by air replacement, and leaving a door partly open quickly proves disastrous for the inside temperature level. In general, users' low sensitivity to this problem, and rapid ageing of draught excluders on doors and windows, give rise to relatively high losses.

Hence in draughty houses it is preferable to use radiation methods of heat exchange (greenhouse-effect wall for example) rather than convection methods (Trombe wall or ventilated greenhouse-effect wall).

Insulation of opaque surfaces

The east, west and north walls (for northern hemisphere) and south walls (for southern hemisphere) are insulated so as to limit losses through conduction. The roof must not be forgotten.

Effective insulation materials are seldom available or they are too expensive to be considered. However, there are appropriate solutions.

For example, cavity walling can be used for the east and west walls, sandwiching air between the outer and inner partitions. This air then acts as an insulator (λ air = 0.024 W/m^2/°C). Air thickness of about 10cm is commonly used. A greater thickness of air would increase convection currents which would result in reduced insulation performance. Banking earth up against outside walls improves their insulation.

Suggesting insulation techniques using local materials is a difficult issue, because these materials are not very resistant to damp and attack by insects and other vermin. Results obtained up to now have been inconclusive.

For areas with least exposure to the sun (north or south depending on the hemisphere) it is advisable to use a buffer zone, which effectively insulates the building from the outside.

Use of local materials for the roof is recommended to ensure high thermal inertia: roofs made of thatch, earth and straw etc. Roofs which lack inertia or which are poorly insulated have a very marked effect on maintaining a high temperature level. Thus corrugated iron roofs are to be avoided at all costs.

Adding a false ceiling of insulating cloth (stuffed with wool or straw) is a simple, effective technique.

The presence of a grain store is shown to play the same role as a buffer zone. Grain stores full of oats, for example, provide very good insulation and warrant a lighter roof.

Figure 83 is an example of a model of direct heat gain for a school, which was designed in a region of Ladakh by the TARA[*] Association.

Comments Double-glazed skylights could be preferable for the buffer zone, so as to provide lighting and heating, rather than single-glazed windows on the northern side, which cause heat loss even during the day.

One could add that the optimum outcome also depends on climatic conditions during the heating period. It is in fact possible to counteract defective insulation with larger-size collector/storage space. However, in the case of short sunny periods or cloudy or misty climates it will, on the contrary, be necessary to reduce collector/storage size (area of glazing,

---

[*] TARA, Technology and Action for Rural Advancement, an Indian association with its headquarters within Development Alternatives, New Delhi, India.

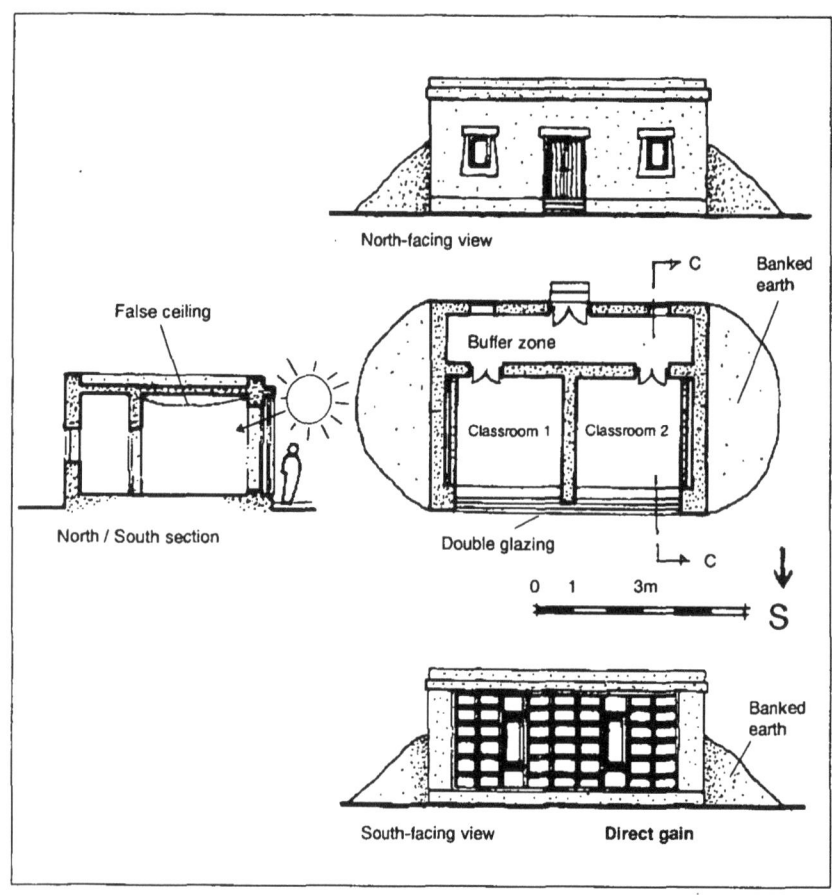

Figure 83 Earth construction primary school
From: Solar Architecture and Earth Construction in Northwest Himalaya

thickness of the interior walls) in favour of insulation, because extra inertia serves no purpose and cannot be used.

Insulation increases the efficiency of passive solar systems. An efficient passive solar system achieves maximum conservation of energy entering it. Marginal internal sources, such as heat released from cooking, metabolism of the occupants etc., help to create a marked increase in the level of heat.

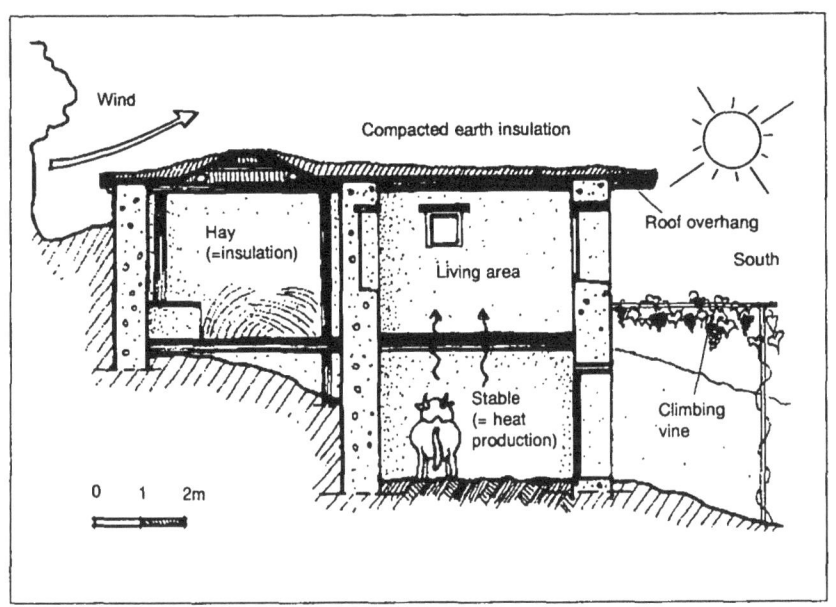

Wind

Compacted earth insulation

Roof overhang

South

Hay
(=insulation)

Living area

Stable
(= heat
production)

Climbing
vine

0   1   2m

Figure 84 Example of rural dwelling (Armenia)

117

9 Building design and choice of system

The design of a passive solar building must meet two requirements for improved performance: rooms to be heated must be positioned on the side of the building which is most exposed to the sun, and surfaces which are in the shade, or exposed to severe conditions (cold winds etc.) must be kept to a minimum.

To respect the first condition it is better to align the building on an east-west axis and provide several levels.

As overall shape of the building determines the heat exchange with the exterior, it is also important to minimize the area/volume ratio so as to limit losses. This explains the compact shape of buildings in cold areas.

Other more concrete factors also come into play: knowing whether a building is old or new, knowing local architectural practices, being innovative with local construction techniques, availability of materials etc.

Before addressing architectural considerations, the designer already has a basic choice of three methods for providing adequate indoor heating of buildings:

o direct heat gain with a glazed area between building and exterior;
o greenhouse-effect wall;
o attached greenhouse.

Details of each system are given in the second part of this book. However, combining two or even three methods can more effectively meet the required objective.

For a room in constant use (living room, bed-sitting room, cattleshed etc.) direct heat gain during the sunny period, together with greenhouse-effect energy for evenings, gives extended cover at times when heating is required.

As for the energy efficiency (ratio of energy recovered by solar heating system to incident energy in the building to be heated), the following hierarchy is established:

o direct gain: 60 to 75 per cent
o collector/storage wall: 30 to 45 per cent
o adjoining greenhouse: 15 to 30 per cent

In temperate latitudes, one square metre of vertical south-facing glazing gives the same quantity of heat as $2m^2$ of collector/storage or $3m^2$ of dividing wall of an attached greenhouse. E. Mazria (see Bibliography) therefore advises that this proportion be retained when methods are combined.

For example:

$5m^2$ of south-facing glazing:
> = $2.5m^2$ of south-facing glazing + $5m^2$ of collector/storage wall
>
> = $2.5m^2$ of south-facing glazing + $7.5m^2$ of partition wall with adjoining greenhouse

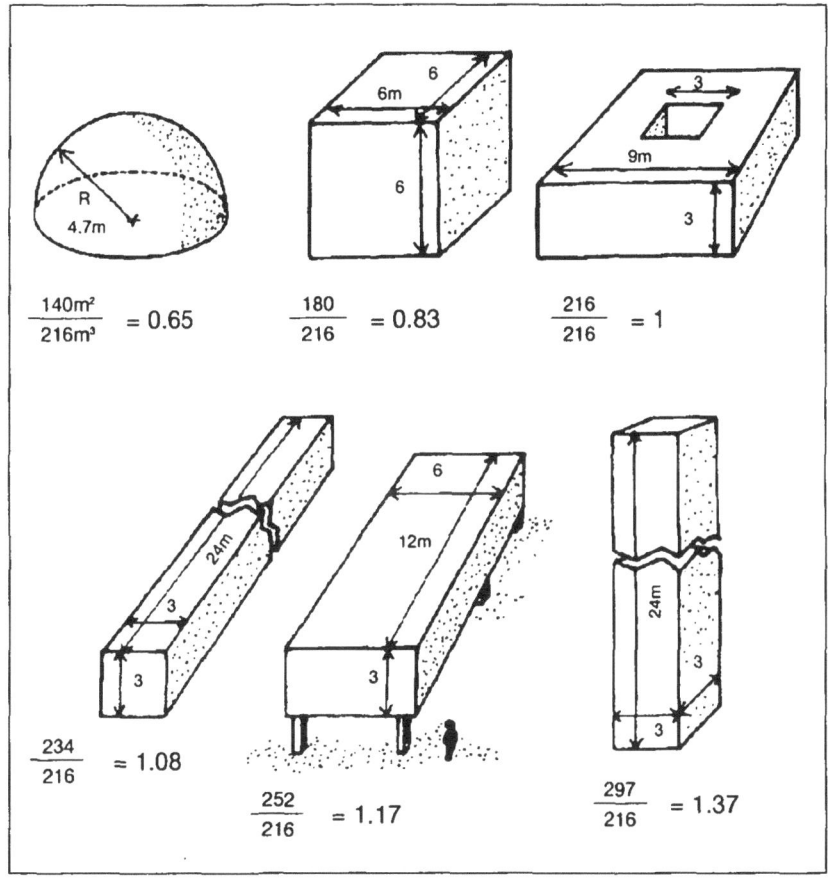

Figure 85 Various geometric shapes determining area/volume ratio geometries

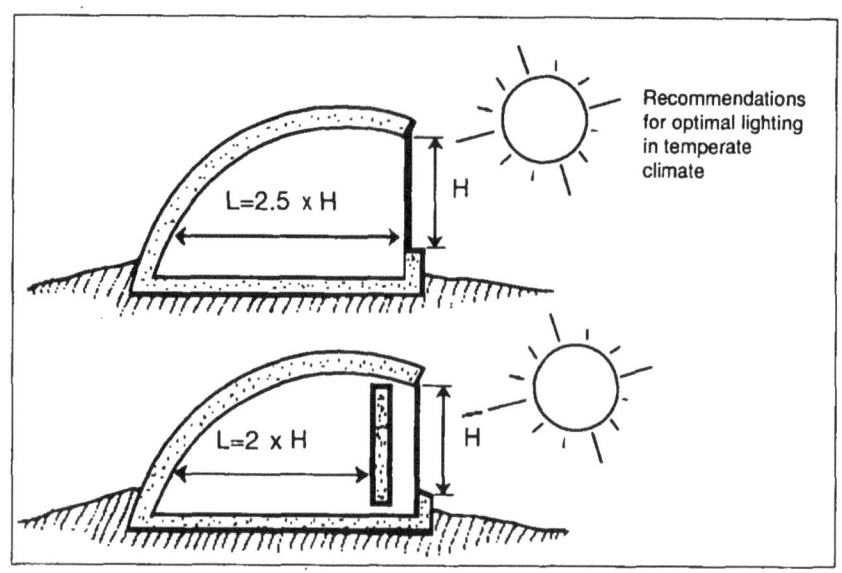

Figure 86 Recommended sizes for attached rooms

From work by J.D. Balcomb for standardized conditions (insulation, sizes, materials)

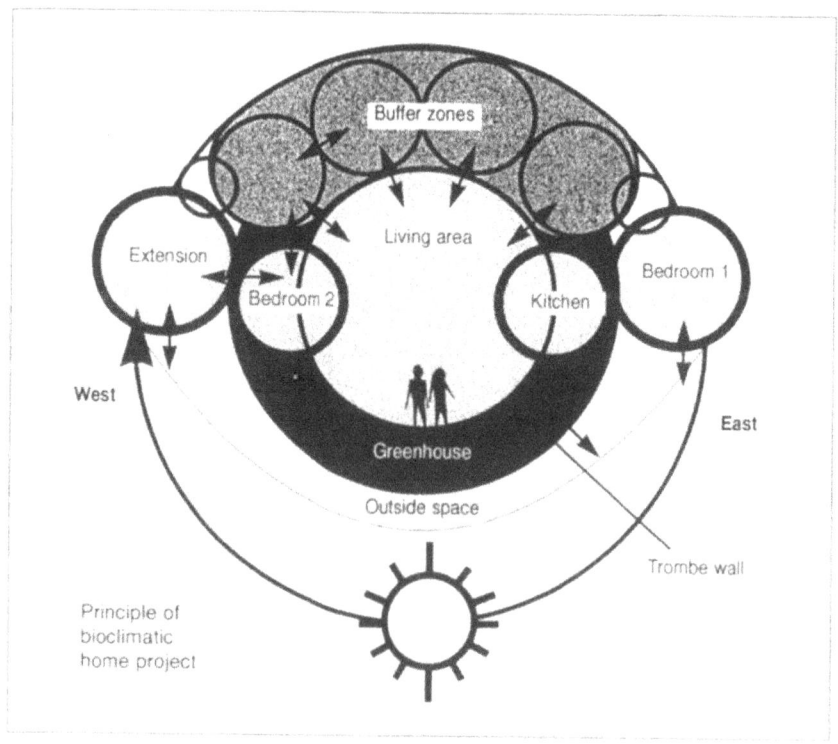

Figure 87 Bioclimatic house facing the sun

10 Predicting the thermal performance of solar buildings

Given the great number of possible parameters relating to location and local construction techniques, there are no immutable rules for sizing buildings. The following methodology is used most often because it is the quickest.

For heating during the cold season, average and maximum variation in inside temperature on a clear day must be calculated so as to predict the thermal behaviour of a building. This system is then adapted according to the particular local climate (frequent cloud cover, for example).

Calculations made for the warm season enable anticipated scenarios to be checked and overheating to be avoided (ventilation, masking). It is also possible to model the behaviour of a building over several days or over a whole year.

To summarize, there are as many methods as there are research institutes, laboratories and architectural practices which have developed a modelling code for their work.

This makes a wide choice of prediction methods available. As for our guide, experimental research enables us to propose two kinds of approaches to the designer: simplified 'manual' methods and microcomputer-based methods.

Simplified manual methods

Most manual methods were developed from results derived from complex computer-based methods. Detailed modelling of physical properties actually requires computers capable of complex calculations. Programs were validated by measurements on experimental installations.

Accuracy depends on the variability of the climate, the behaviour of users and on so many unpredictable parameters, even in the very sophisticated models. In any one-month period, a variation of six to eight per cent is to be expected between the thermal behaviour of a building and the behaviour predicted by crude calculation.

These methods, developed in the USA and Europe, are based on statistical treatment of climate data for a minimum period of 10 years. But in the majority of cold regions in developing countries it is currently impossible to obtain statistics for such a long period. High-altitude meteorological stations do not in fact possess sufficient solar energy measuring equipment.

Each country moreover has distinctive architectural features and its own particular conventions. A simplified method therefore has to be adapted in each case to respect any existing standards.

Here we present the method developed by Edward Mazria in 1979. Its major appeal lies in its simplicity, which is also its limitation, particularly when buildings become complex. Nevertheless it gives an accurate prediction of system efficiency.

This method has five stages.

Stage 1: Calculating heat loss in winter

If the system has to achieve worthwhile results in the coldest season (for livestock, plant growing, public buildings etc.), a representative clear day in the coldest month is selected. Otherwise it is quite possible to choose a different reference period.

Losses through each surface (walls, glazed area, ceiling, floor)

d = K.A

    d = losses in W/°C
    K = conductivity of the area in W/m²/°C
    A = surface area in m²

Let us consider the particular case of heat loss through the floor.

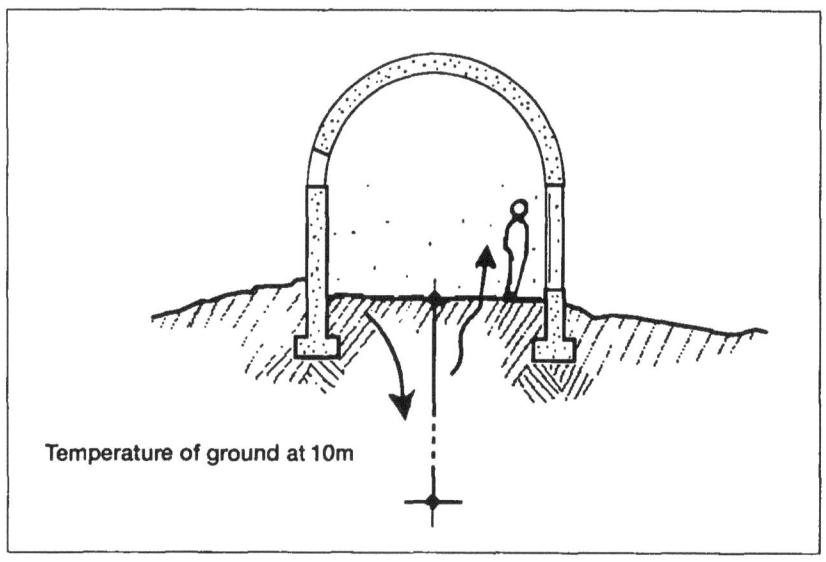

Figure 88 Heat loss through floor

122

Table 11: Summary calculation table

Element	A (m²)	x	K (W/m²/°C) =
Front walls	x		=
Roof	x		=
Floor	x	0.5	=
Door	x		=
Windows	x		=
Replacement of air		x N x 0.34 (Wh/m²/°C)	=
		d total	= w/°C)
(and afterwards calculation of F × (W/m² floor/°C)			

According to convention, heat exchange through the floor is defined such that it has a temperature at ten metres depth which is equal to the average annual temperature for the location in question.

With thermal conductivity λ for earth of the order of 2W/m/°C gives $K_{EARTH\ FLOOR} = 0.2W/m^2/°C$ (for the 10m-deep layer)

Losses from air replacement

The formula: $d_{replacement} = 0.34.N.V$ (in W/°C) is used.

N: hourly rate of air replacement;   V: interior volume in m^3

We have:

$$d_{total} = d_{area} + d_{replacement}$$

In short, to provide a comparative method, this loss in capacity is reduced to an overall loss coefficient (F), expressed in watt-hours per day per m^2 of floor and per degree such that:

$$F = \frac{D_{total}}{floor\ area}\ 24\ hours\ (Wh/day/m^2\ /°C)$$

Stage 2: Calculating solar energy supplied

At first, solar energy is measured for all glazed areas – those which let light into the interior and those positioned in front of a connecting wall.

There is a choice of two methods:

o The graphic method, which was proposed above, allows solar contribution for a type of glazing to be calculated by superimposing a solar diagram for the month in question on the radiation indicator according to the angle of inclination of the glazed surface.

To use diagrams provided in all circumstances a correction factor must be applied to take account of the transmission rate of the glass used and of the ratio of the transparent area to glazed area given in the table.

123

o The direct method according to climatic data (ideal if such data is available).

There are direct contributions :

$$DG = A.I.\text{correction factor}$$

DG: direct solar contribution (direct gain)
A: area of glass not in the shade in m^2
I: solar contribution per m^2 of glass in Wh/day

and indirect contributions:

$$IG = A.I.p$$

IG: indirect solar contribution (indirect gain)
A: area of glass not shaded (in m^2)
I: solar contribution per m^2 of glass in Wh/day
p: percentage of incident energy on a collector wall which reaches the interior

Mazria has chosen to express 'p' by the ratio of collector area to the floor area of the room to be heated (see Figure 89).

The percentage of transmitted energy or efficiency of the collector wall is a function of the quality of the insulation of the building to be heated.

The masonry storage wall provides an example (30cm thick, double glazing, black exterior surface).

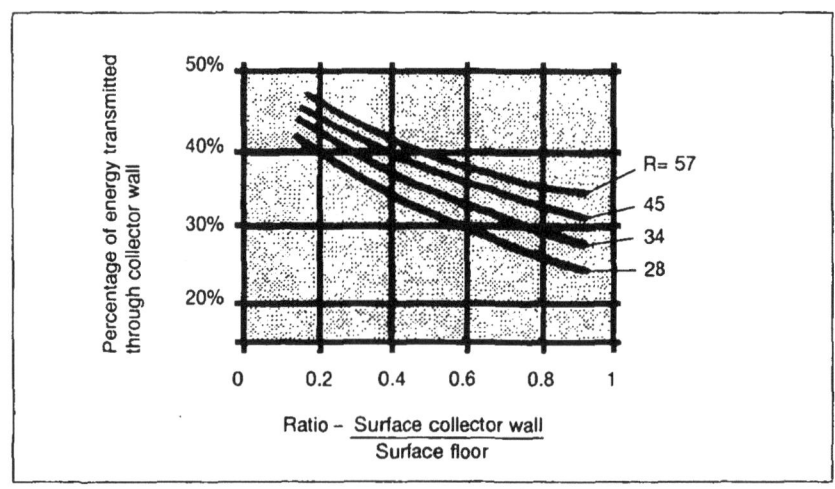

Figure 89 Percentage of energy transmitted through 30cm-thick masonry wall

124

For an initial approach to systems using a greenhouse-effect solar collector greenhouse wall (Part Two) choose: p = 25 per cent. For party walls with adjoining greenhouse: p = 15 per cent. Subsequently a total thermal contribution is defined per square metre of floor space (C in Wh/m^2 of floor):

$$C = DG/ \text{floor area} + IG/ \text{area of floor}$$

Stage 3: Determining the average indoor temperature

Assuming that the average temperature has reached an equilibrium after a series of identical days (solar radiation, outside temperature, wind etc.), it is not necessary to take into account the role of the building's inertia, as the average temperature is then only a function of heat entering the building and losses incurred because of its design.

Initially no account is taken of the internal contributions from stoves, inhabitants, cooking etc. The internal temperature is given by:

$$t_i = C/F + t_0$$

C: coefficient of thermal contribution in $Wh/day/m^2$
F: coefficient of thermal losses in $Wh/day/m^2$ of floor/°C
t_0: average daily outside temperature in °C

Stage 4: Determining daily variations in indoor temperature

This stage is important because it gives an idea of the way the building responds under dynamic conditions. It also indicates the time at which a particular process makes the most significant contribution to the ambient temperature of the building.

For the direct gain methods presented in Part Two it is possible to draw a theoretical curve of indoor temperature variation of the building, taking into account the characteristics of traditional architecture in cold regions (massive earth walls, poor insulation, heavy roof, low ceilings etc.) (Figure 90).

For direct gain it is difficult to calculate the curves of temperature variation because of the many different factors involved.

A method developed in 1984 by the Los Alamos National Laboratory (LASL, New Mexico) can be quoted. This takes account of all internal parts of the building and distinguishes between contributions made by radiation and convection.

Defined as diurnal heat capacity, this method is called *DHC* for each storage and non-storage inert area according to its own characteristics during the day, from 06.00 until 18.00 (solar time). The quantity of solar

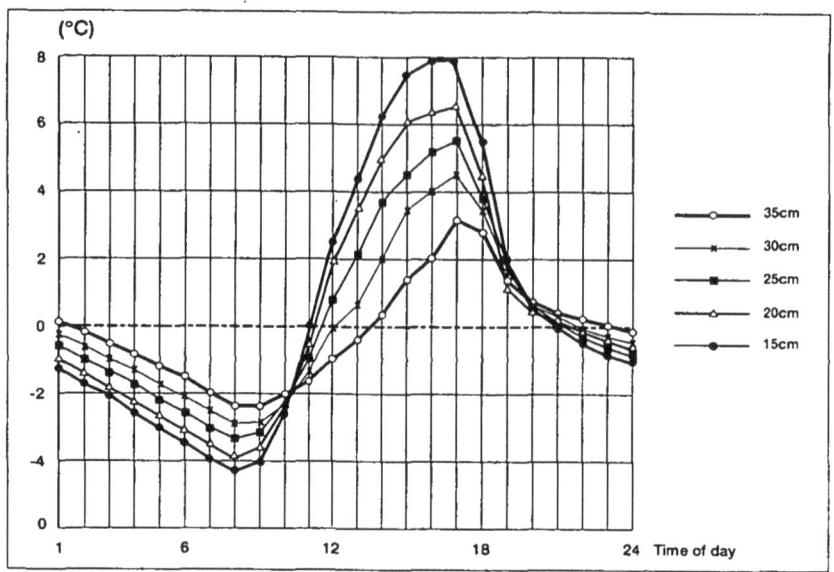

Figure 90 Indoor temperature variation of buildings with collector/ storage walls or connected to a greenhouse by an adjoining wall

energy stored and the corresponding lag for each area are thus taken into account and allow variations in indoor temperature to be described accurately.

The controlling equation for a building heated 100 per cent by solar energy is as follows:

$$\Delta T = 0.61 \cdot Qs/DHC$$

o ΔT: the amplitude of variation in indoor temperature over 24 hours
o Qs: heat transmitted by solar methods
o DHC: the sum of A_i and dhc_i, A_i being the area of storage surface and dhc_i their respective diurnal heat capacities.

The Los Alamos Laboratory has a computer program to facilitate calculation of the DHCs. For the development of the calculations please refer to the book cited below.[*]

[*] *The method of calculating total DHCs for a building is given in :* Heat Storage and Distribution inside Passive Solar Buildings. *Los Alamos National Laboratory Report LA-9694 MS 83 and 'Predicting of Internal Temperature Swings on Direct Gain Passive Solar Buildings', Los Alamos National Laboratory International Report LA UR 83.2246. Available at GERES.*

Stage 5: Top-up heating requirements

This calculation must be made for buildings which need to maintain a constant inside temperature: operating theatres, buildings used for agricultural production, etc.

First of all the heating requirements of a building have to be determined in terms of a fixed target temperature and differences in temperature as compared with the exterior. To make a good approximation, calculations are made for a monthly period.

Thus $Q = F \cdot Af \cdot Ddm$

Q: monthly heating needs in kWh (kilowatt hour)
F: coefficient of losses from building in kWh/day/ $m^2/°C$
Af: floor area in m^2
Ddm: degree-days for the month

For a single day, the number of degree-days is the difference between the average outside temperature and the target temperature.

For a month, the number of degree-days for the different days of the month is totalled.

The top-up heating needed then remains to be calculated:

$$Q_{\text{top-up heating}} = Q - Q_{\text{solar}}$$

Q: monthly heating requirement in kWh

Q_{solar} : monthly contribution from solar heating in kWh (see above)

$Q_{\text{top-up heating}}$: top-up heating requirements in kWh

The method proposed by the Los Alamos Laboratory provides charts for typical passive solar system layouts (veranda, ventilated and unventilated Trombe wall etc.) which indicate the fraction of heating saved by the solar system (in comparison with the same construction under this system, when the parts are replaced with perfect insulation preventing all thermal exchange with the exterior) according to climate and the special characteristics of local architecture.

The advantage of the method lies in its universal nature. It has therefore been tested in a great variety of climates. The correlation curves were drawn up with coarse computer code, the Pasole program. In France it was adapted in the name of the B-SOL method. It is also used in China. (See Bibliography for further information.)

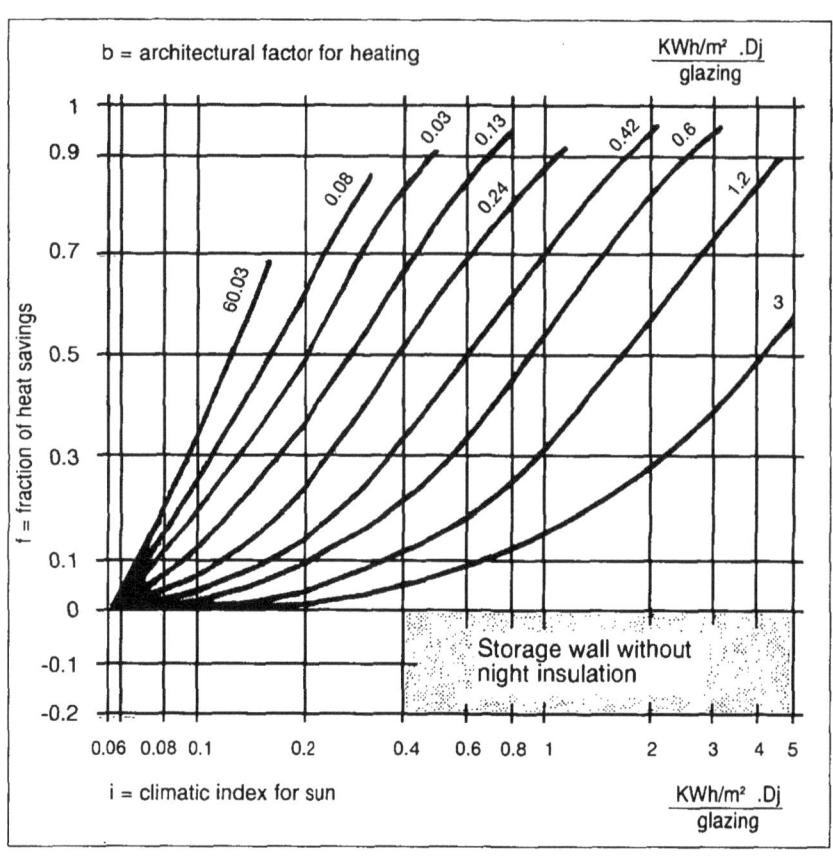

Figure 91 Example of chart for a solar wall
(from E. Mazria)

Computer methods

There is a very wide range of software for thermal simulation adapted to housing conditions, varying from the complex codes requiring large capacity computers to simplified methods of analysis using a micro-computer.

All operating software makes use of the following data:

o definition of exterior climate (temperature, insolation, wind, humidity etc.);

o details of building (materials, walls, glazing, heating systems, orientation etc.);

o details of environment (possible shading, altitude, longitude and latitude etc.).

128

Depending on the individual case, results of simulation can provide heating requirements of the building for a given target temperature, and progress of temperature over time, for a given area.

Figure 92 shows several programs adapted for cold regions.

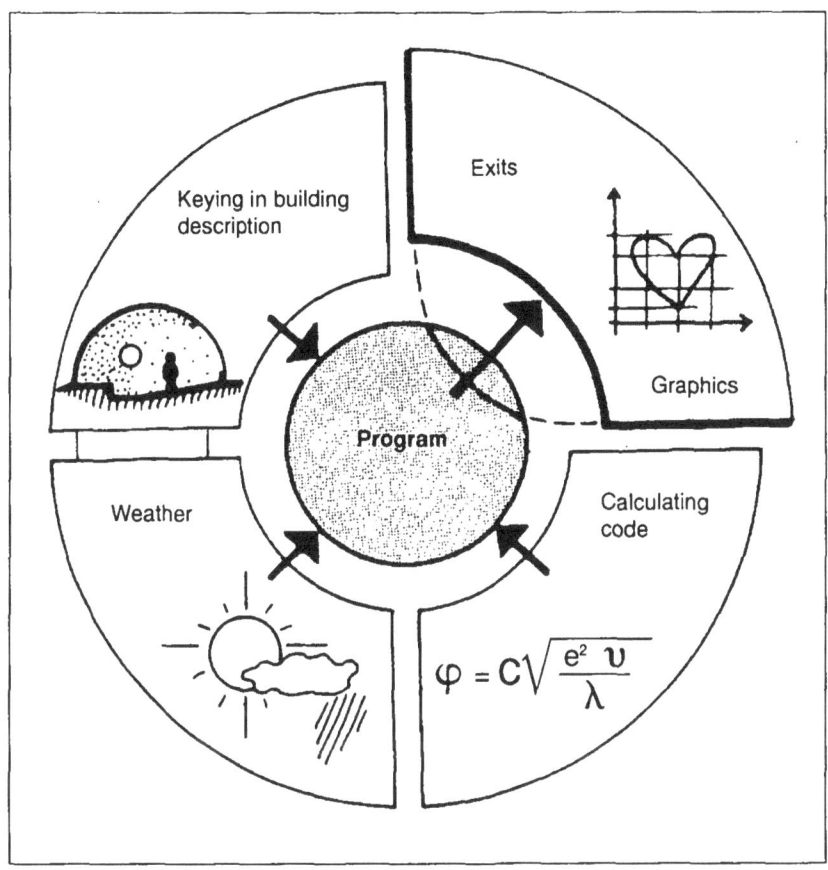

Figure 92 Principle for operation of thermal simulation program

Table 12: Programs for calculation of heating needs and thermal behaviour of buildings

Title	Designer/distributor	Function	Codes used	Cost (1993)	Comments
COMFIE	Ecole des Mines, Paris B Peuportier 60, Bd Saint-Michel 75272 Paris Cedex	Calculates dynamic heating requirements, zone by zone (maximum 10) Disseminates area temperature profiles	Nodal networks, modal analysis, works on IBM AT or compatible and Macintosh SE/II	FF500	Short learning period
SIMULA	IPTIC Maison de l'Ingénieur 3, rue Léon Bonnet 75016 Paris, France	Simulation in varied conditions of a multi-zone building	Compatible with IBM AT	FF10 000	Training needed
MICROPAS	B.E. Adret 2, rue Clovis Hugues 05200 Embrun France	Disseminates temperature profiles	Nodal Networks, no hourly time, French version of the American software CALPAS 1 (5 zones)	FF35000	Training needed
TRNSYS	TRNSYS Coord. Solar Energy Lab 11303 Erg Res Buildg. University of Wisconsin 1500 Johnson Drive Madison, WI 53706 USA	Graphical output of temperature profiles Calculates hot water systems	Response factors, nodal networks, high level code IBMPC/AT compatible	US$800	Customized training needed

(continued overleaf)

Table 12 (contd.)

Title	Designer/distributor	Function	Codes used	Cost (1993)	Comments
MICRO DOE2	Acrosoft International 9745 East Hampden Ave Denver, Colorado, CO 80231 USA	Calculates heating requirements Graphical output	Response factors, weighting, high level code	FF6065	Training needed
TAS	OASIIS - Parc Ste Victoire Bât.2 - 13590 Meyreuil France	Output graph of losses Air speed Active systems	Impulse method (response factors), Apollo HP station	(With Apollo HP station) FF400 000	Training needed

131

APPENDICES

Principles for use of a solar greenhouse

The principles of use are presented in note form. These notes are drawn from the experience of GERES in the Indian Himalayas, from the work of B. Verne.

Note 1 : control of temperature under glass

Too high a temperature under glass poses as many problems as a sudden frost. If the greenhouse is only planted with traditional vegetables, that is mangal (a type of beet), palak (a local variety of spinach) or lettuce, a temperature of around 18° to 20°C is sufficient at the time when germination is at a high level. The temperature should never exceed 25°C.

For cabbages the temperature can be limited to 15°C. For tomatoes and courgettes it is better to have a temperature between 22° and 25°C.

Once the temperature is too high in the greenhouse as much air as possible should be circulated through the greenhouse by opening the door and the aeration holes, or by partly removing the polyethylene covering so as to restore 'normal' temperatures.

It is also very important to limit the differences in temperature between day and night. In Ladakh the only solution which can be adopted is to cover the greenhouse during the night, with blankets for example.

The greenhouse must be covered 30 to 45 minutes before sunset and uncovered after sunrise.

Controlling the temperature of the greenhouse ensures:

o limiting development of diseases and parasites and especially greenfly attacks
o easier growth and development of vegetables under glass
o 'strong' seedlings produced for planting out: i.e. they will survive planting out without problems.

Note 2: soil sterilization

The soil in the greenhouse has to be sterilized approximately every two years to avoid the development of certain diseases and parasites.

The simplest method is to heat up a metal plate, place the earth to be sterilized on it and then spray to keep the soil moist. Ten to fifteen minutes after steam appears are enough.

Also, in spring, the first fifteen centimetres of soil can be replaced by fresh soil from outside. This operation, repeated every two or three years, prevents soil depletion and limits the development of diseases and parasites.

Note 3 : rotation

Growing the same species of vegetables for most of the year, on the same shelves, contributes to the development of diseases and parasites and soil depletion. This problem is limited by encouraging rotation. There are several alternative approaches, and they can be combined:

o alternating leafy vegetables (lettuce and spinach) with root vegetables (turnips, spinach beet, beetroot family) every two years. If possible, vegetables needing strong manure are alternated with those which require less; for example chinese cabbage, spinach (vegetables with a high-level requirement) with turnips (lower requirement) and lettuces and onions (reasonable requirements).
o allowing the soil to rest for a month or two, for example in the coldest winter month (January). In such a case, a system of rotation can also be designed, which allows each shelf to rest for one month in turn. For a greenhouse with six shelves each shelf will thus 'rest' twice in a year.
o introducing a non-market-gardening crop such as peas in the summer improves soil fertility and reduces the risk of disease and parasites. Peas can be used as green manure if harvested before they are ripe.

Schematic example of rotation for a greenhouse of six shelves of 150cm × 120cm:

I	II	III

VI	V	IV

First year
Autumn/winter

o Chinese cabbage on shelves I and II.
o Spinach on shelf III.
o On shelves IV and V: two consecutive sowings of coriander. Then a possible sowing of lettuce or turnips. Provide a one-month rest period in winter.

o On shelf VI: start with salads and/or rdums (leaf vegetables), followed by a possible sowing of coriander, then a rest period of four to six weeks.

Spring/Summer

o One-month rest on shelves I and II, after mangal, then pea crop and rest in autumn (the peas can be harvested green for animal feed).
o On III: rest for a month, then courgettes or aubergines.
o On IV and V: sowing for planting out cabbage and tomatoes, then manure and summer rest.
o On VI: sowing for onion plants, then summer rest.

Second year
Winter

o On I : spinach.
o On II: two sowings of coriander, possibly followed by growing turnips or salads, then rest.
o On III: the same.
o On IV: salads.
o On V and VI: spinach; beet.

Note 4 : making compost

Making compost in cold regions of developing countries is subject to two major constraints: availability of materials and climatic conditions.

The materials available are:

o essentially dead leaves
o kitchen waste: rice and vegetables when they are not used to feed animals
o animal waste and human faeces
o garden waste, when it is not used for animal feed
o wood chips from carpentry workshops (to provide the carbon content).

For leaf-based compost it is better to use different species and leaves which are still damp after falling. They should therefore be collected very quickly.

Climatic conditions
Because of the very long harsh winter, composting can only be achieved using a solar greenhouse, preferably a polyethylene one (the same compost corner can be used in summer by lifting the glass cover).

Making compost If possible try to have a heap of compost of about two cubic metres in volume, 80cm in height and 120 to 150cm in length. In every case the compost heap must be sufficiently dense. To achieve sufficient aeration, compost which is mostly made up of leaves should not be compressed.

Composting is carried out in various phases In summer, all possible materials are stocked in a place protected from the sun (which would limit fermenting), next to the composting space.

In autumn, a heap of compost is made as soon as the leaves fall, mixing them in with materials collected during the summer. A certain quantity of traditional manure can be added in order to start the process of fermentation by providing nitrogen. When the heap has reached its final size (approximately 150cm × 120cm × 80cm), it is covered with several (5 to 10) centimetres of earth. Every 4 to 6 weeks the heap is carefully turned and mixed and efforts are made to achieve, little by little, the most homogeneous mixture possible.

The interior of the heap of compost should have a temperature of about 60°C, with humidity between 40 and 60 per cent.

If the temperature seems too low, nitrogen is added in the form of kitchen waste, vegetable waste or traditional fertilizers – excrement – in order to reactivate the fermentation process.

If the temperature is too high or if the compost seems too dry, the heap should be dampened with a watering can.

The temperature under glass should not be excessive (no more than 35°C during the day). Night-time temperature falls are avoided by covering the greenhouse at night.

A good compost does not smell bad: this is a reliable indication that fermentation is progressing well, as are the temperature and humidity inside the compost heap.

Note 5: data on planting density

Densities given are representative and are based on those used in France.

As climatic conditions in cold regions are very localized and local species are robust, it is wise to carry out several trials on density around the values given below in order to find the best ratio between the number of plants per square metre and the quantity of leaves (or roots) harvested for each plant.

For chinese cabbage: use a basic spacing of 30cm between plants in a row, with 35 to 40cm between rows.
Spinach: sow 'densely' in rows, 20 to 25cm apart.
Lettuce: If sowing is too dense do not hesitate to thin them by removing a certain number once they reach the stage of having three or four leaves.

When sowing in rows keep 10 to 20cm between plants if possible, with row spacing of 25 to 30cm.

Turnips: Sow in rows, with spacing of 20cm between rows and 10cm between plants. Broadcast sowing, if done not too densely, can also be used.

Because of the size of the seeds it is difficult to control the quantities sown. Thinning should therefore be carried out when plants reach the stage of having three or four leaves.

Note 6: growing tomatoes and cabbage seedlings

When growing for planting out it is very important to control effectively:

o greenhouse temperature (see Note 1)
o density of sowing
o planting out dates

If these three factors are managed, growing seedlings should not pose problems.

Tomatoes/aubergines/courgette seedlings

Tomatoes are demanding plants: they require a constant temperature of 22 to 25°C during seed germination.

When plants have reached the fully extended 'cotyledon' stage (cotyledons are the tiny leaves which appear at germination and disappear afterwards), the temperature can gradually be lowered to between 10 and 12°C.

At the time of germination the shelves sowed with tomatoes are covered with plastic at night or on cold days.

It is very important not to sow too densely.

The higher the sowing density the less sturdy the seedlings will be and the less they will survive planting out.

The ideal is to have a maximum sowing density of 100 to 150 seedlings per square metre, with rows spaced about 10cm apart.

Seedlings can be planted outside between 30 and 50 days after sowing.

When planting out the night-time temperatures must be high enough (at least 10–12°C) if they are to grow in the best condition.

It is best to make several sowings with spaces of a week or ten days between them.

When planting out outside, spacing is about 30cm between plants (about one foot) and 80cm between rows to give a density of six plants per square metre.

The temperature requirements for aubergines and courgettes are the same as for tomatoes.

For aubergines: sowing under glass should not exceed 80 to 100 plants per square metre, with planting out 40 to 50 days after sowing. For planting out the density is two plants per square metre (50cm between plants in a row, one metre between rows).

For courgettes: sowing can be direct, under glass or outside. Seed is first soaked for a day in not too cold water. Sowing density is three plants per two square metres.

Cabbage and cauliflower plants

Cabbage plants are much less demanding in terms of temperature than tomato plants: 15°C is sufficient for two or three days at the time of germination.

Higher densities are possible for the tomato, but as growing conditions are harsh they should not be so dense as to prevent plants from growing 'strong'.

It is best not to have more than 300 to 350 plants per square metre, sowing in rows spaced about 10cm apart.

Planting out occurs a month or a month and a half after sowing. Temperatures of as low as 8°C can then be tolerated.

Practical guide to using a solar hen house

This guide is drawn from the experience of Runamaqui in Peru for the raising of chickens for consumption (see Bibliography).

The hen house is situated at high altitude. Chicks of around 40 grams are brought from low valleys and fattened up to adult size. The chickens are intended for sale.

Preparation of the building

Temperature:
A high temperature is essential at the start of rearing; everything possible must be done to achieve this level (glazing must be cleaned, door closed etc.).

Hygiene:
The building must be completely disinfected. Typical disinfectants are crecyl, phenol, iodine, chlorine and formaldehyde.

Litter of local straw, 5 to 10cm deep is provided.

Arrival of the chicks – first day

Sorting:
Chicks suffer in transport. Dying chicks and those below 25 grams in weight therefore have to be removed.

Feeding:
Suitably ground feed is spread out on paper arranged on the litter.
Fresh sugared water will provide energy for the chicks and help clean their stomachs.
A temperature of 34°C is needed. Top-up heating may be necessary for the newcomers.

First week:
The feed and water troughs are cleaned every day.
Feed is weighed and spread out in the feed trays.
On the third day vitamins are added to the feed and antibiotics added to the water.
The temperature is kept at 32°C.

Second week:
A second type of feeding tray is introduced to match the size of the chicks.

Third week:
The temperature can be reduced to 24°C.
The feeding troughs are suspended.
The litter must be cleaned and 5cm of straw added.

Fourth week:
The temperature is reduced to 20°C. At four weeks the chicken has reached a respectable size (over 450 grams).

Rearing can of course continue. Between six and eight weeks the chicken reaches its maximum weight (2kg).

Logbook
In order to monitor progress in rearing, a daily notebook should be kept to record:

o weight of food distributed
o weight consumed
o weight of chicks
o vaccinations given.

Monitoring equipment

o thermometer
o hygrometer (relative humidity between 60 and 70 per cent)
o weighing scales

Feed for chickens
Normally concentrated conventional chicken rearing uses feed from the large food and agriculture companies. However it is expensive and availability is poor and not well suited to the needs of chicken farming at high altitude. Local production is possible using available nutritional material.

An adequate local supply was used by Runamaqui which gave better results for a lower cost than industrial feed.

Vaccination
Against Newcastle disease (a viral infection affecting respiratory and nervous systems): vaccination is carried out on the ninth day in the eye (a few drops in the eye), to be repeated five weeks later, then every two months.

Against diphtheria, smallpox and plague (viral infections): essential vaccination is carried out on the eighteenth day. When infection is detected, specific antibiotics have to be used.

Radiation indicators and cylindrical solar diagrams

Here the designer will find:

o radiation indicators (very clear sky conditions, cloud factor TL = 2) for horizontal receptive areas (i = 0°), vertical (i = 90°) and sloping (i = 30° and 60°);
o the cylindrical solar diagrams for latitudes from 0° to 50° North and South.

It is advisable to copy the required indicator in advance onto transparent tracing paper and to overlay it onto the solar diagram of the latitude concerned.

The diagrams are drawn from the work of Jean-Louis Izard, researcher-lecturer of the A.B.C. group (see Bibliography).

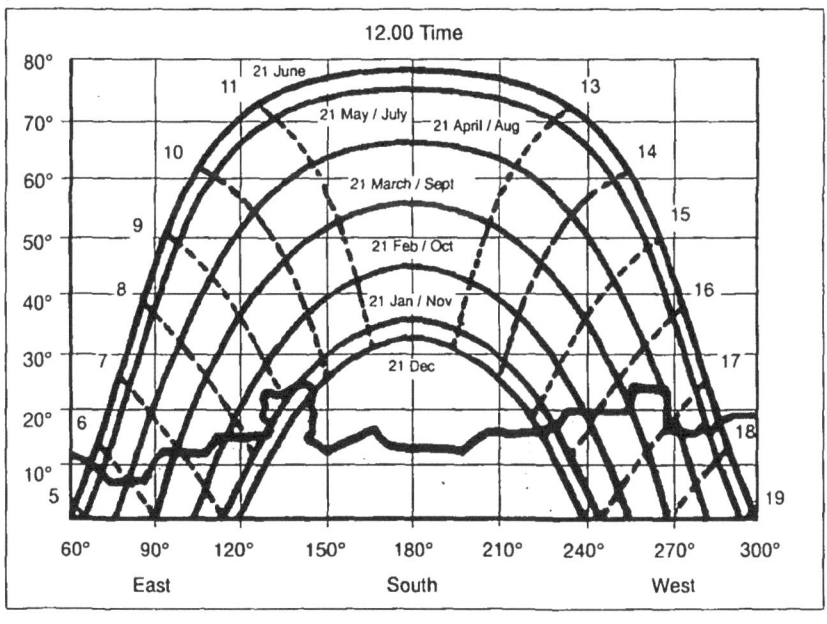

Figure A1 Upper limit of shading on solar diagram

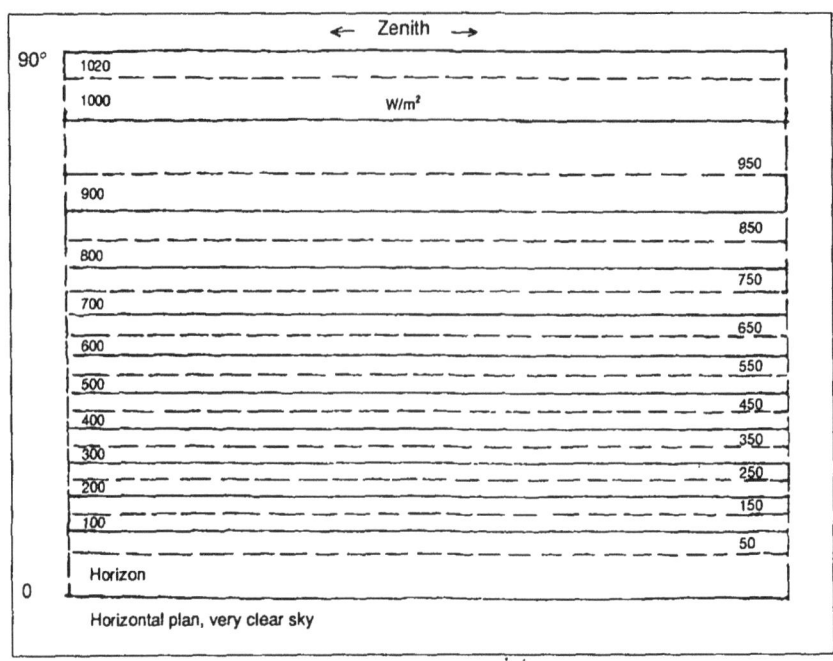

Figure A2 Radiation indicator: horizontal plane, clear sky

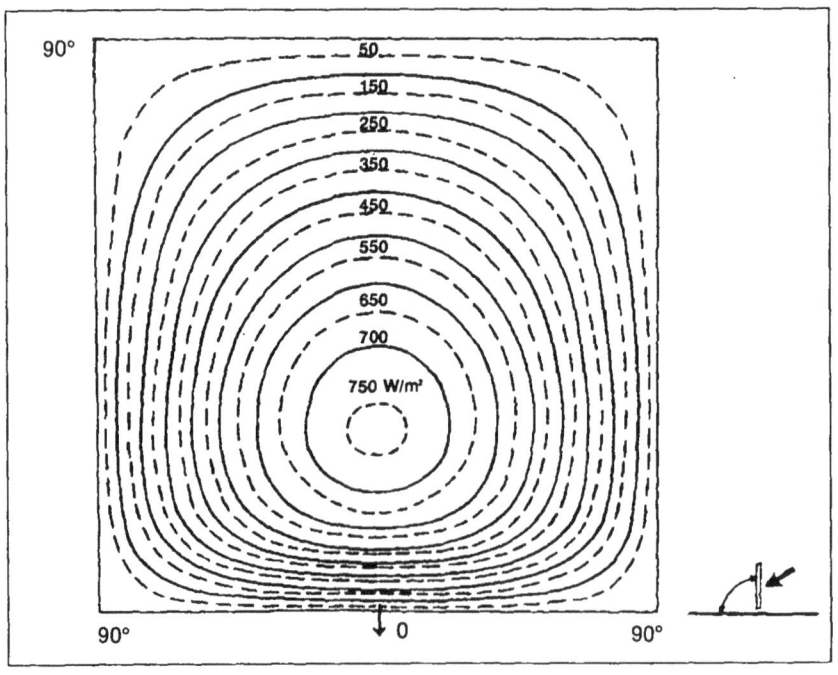

Figure A3 Radiation indicator: vertical plane, very clear sky

144

Taking into account local shading

For a given site possible shading should be clearly marked on the solar diagram. To do this the line indicating the upper limit of the surrounding relief which marks part of the sun's path is drawn on the solar diagram.

Marks are made for the orientation (angle relative to due south) and the height (the angle between its highest point and the horizontal at the location concerned) of each obstacle. Joining these different points gives a line on the solar diagram. Below this line the sun is hidden from the solar collector area. The energy received is thus non-existent and is not taken into account in the calculations.

For the two indicators on the following page (Figures A4 and A5) the curves between $-180°$ and $0°$ can be obtained simply by reproducing the diagrams symmetrically about the $0°$ azimuth on the left-hand side. The share of luminosity for azimuths between $-180°$ and $+180°$ can thus be obtained.

The subsequent solar diagrams apply to latitudes from $0°$ to $50°$.

For latitudes below $23° 27'$, the course of the sun changes hemisphere depending on the time of year. It is therefore necessary to have a solar diagram covering $360°$ to represent the different courses of the sun. Here the diagram for the equator at latitude $0°$ is given (Figure A6).

For other low latitudes only the left side of this diagram for the northern hemisphere and the right side of the diagram for the southern hemisphere is provided. To reproduce the solar diagram for $360°$ the part of the diagram supplied should therefore be traced onto paper and the inverted tracing added to it. Care is needed to respect continuity of the sun's course and times while referring to the diagram provided for latitude $0°$ (Figures A7–A14).

Above this latitude (in reality $23° 27'$), the sun does not pass the zenith and remains in the same hemisphere as the observer. The times and dates for the northern hemisphere diagram are simply reversed to obtain the diagram for the southern hemisphere (values in brackets, Figure A15). Subsequently these values in brackets will not be shown: the same reversal for the southern hemisphere simply has to be copied (Figures A16–A20).

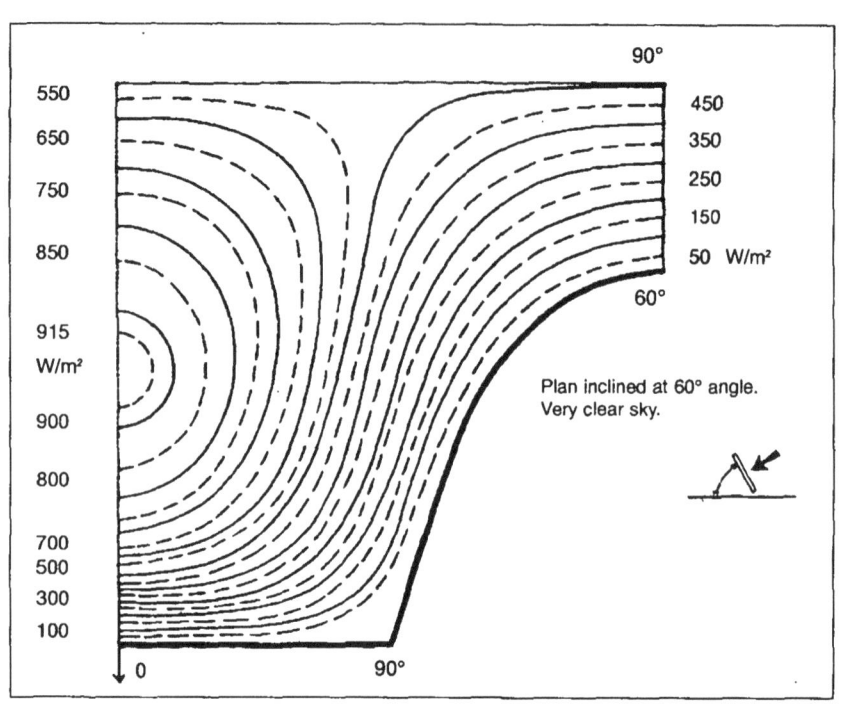

Figure A4 Radiation indicator, 60° angle

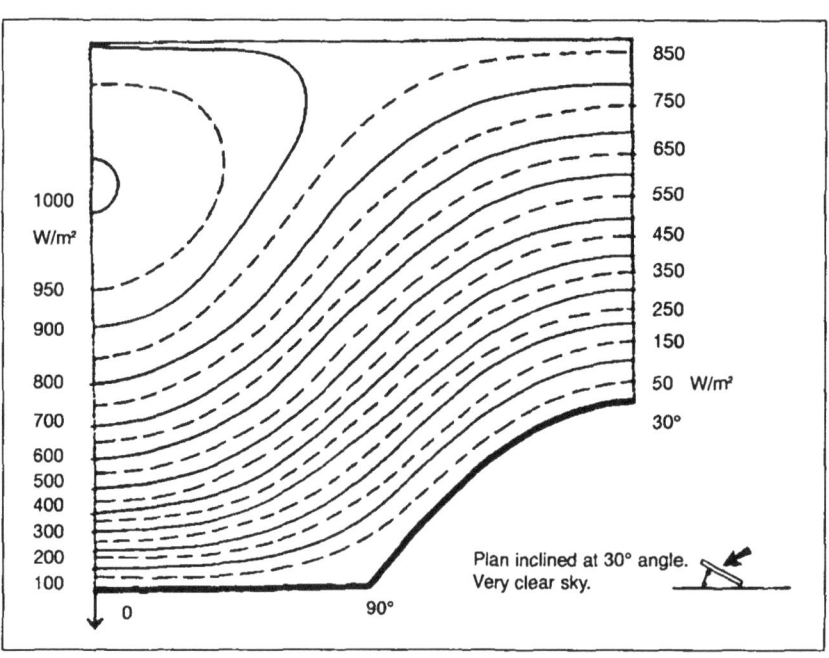

Figure A5 Radiation indicator, 30° angle

146

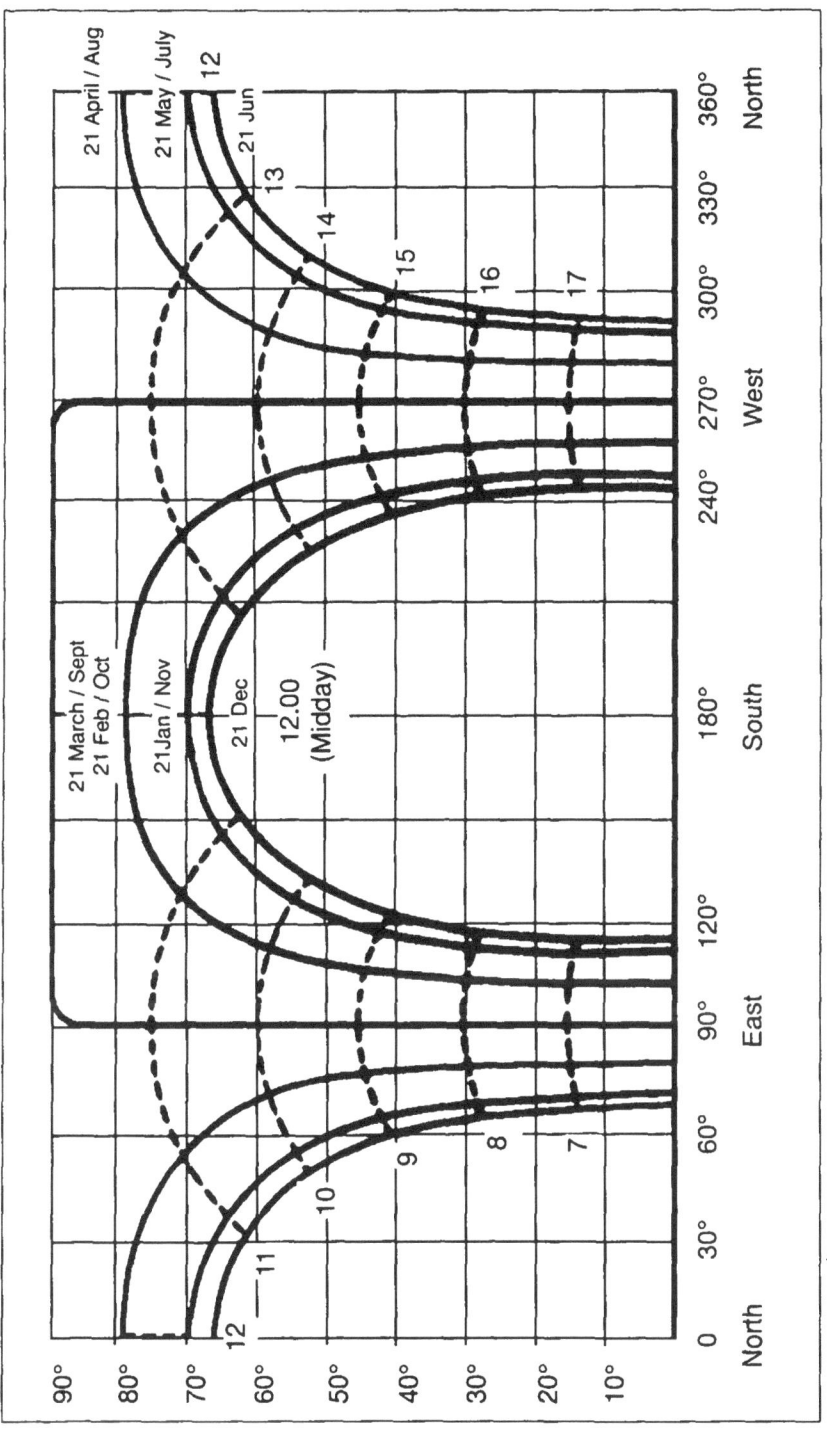

Figure A6 Solar diagram for latitude 0°

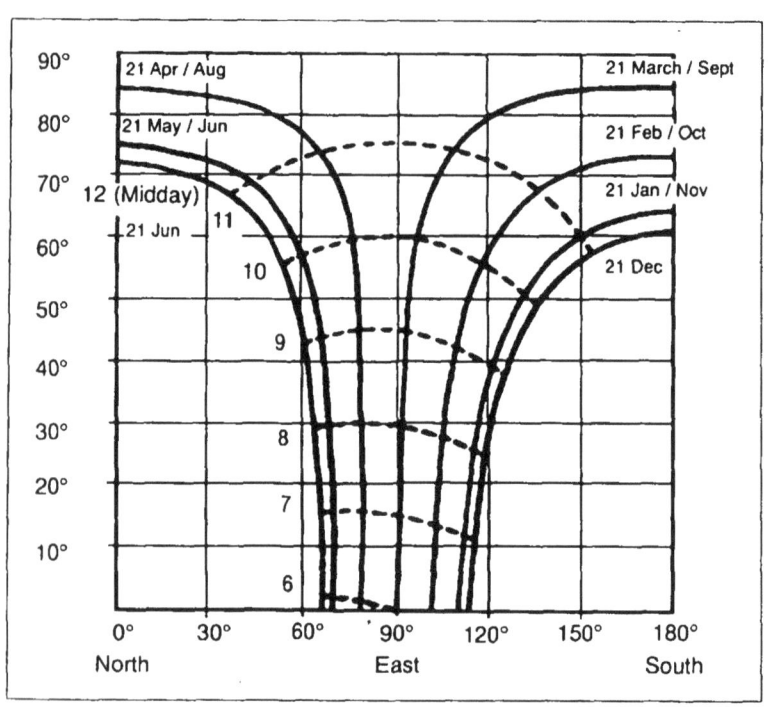

Figure A7 Solar diagram for latitude 5° North

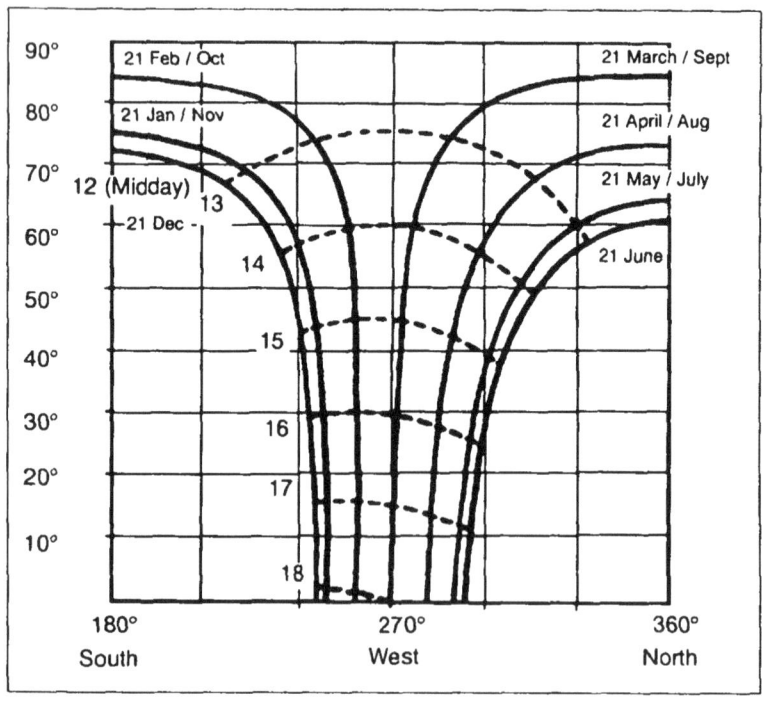

Figure A8 Solar diagram for latitude 5° South

148

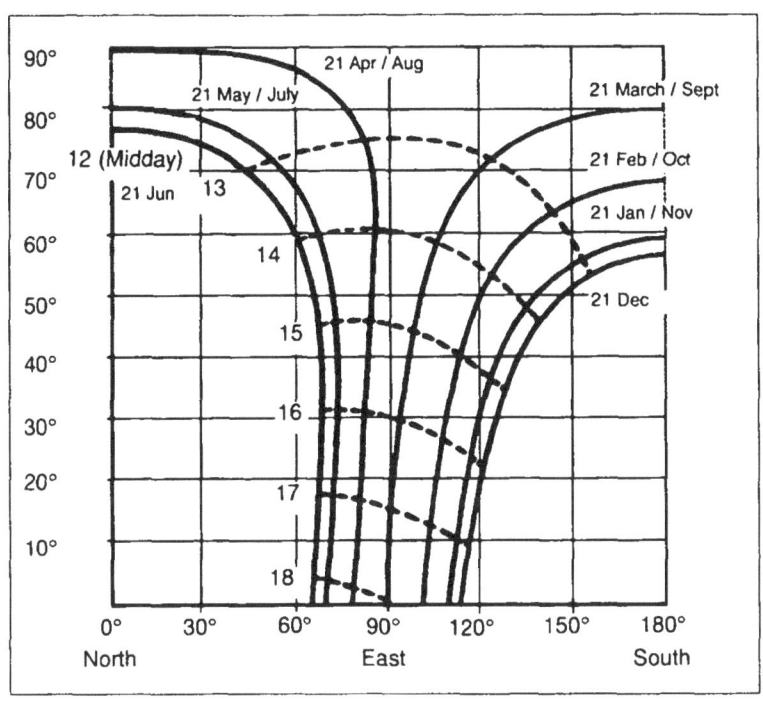

Figure A9 Solar diagram for latitude 10° North

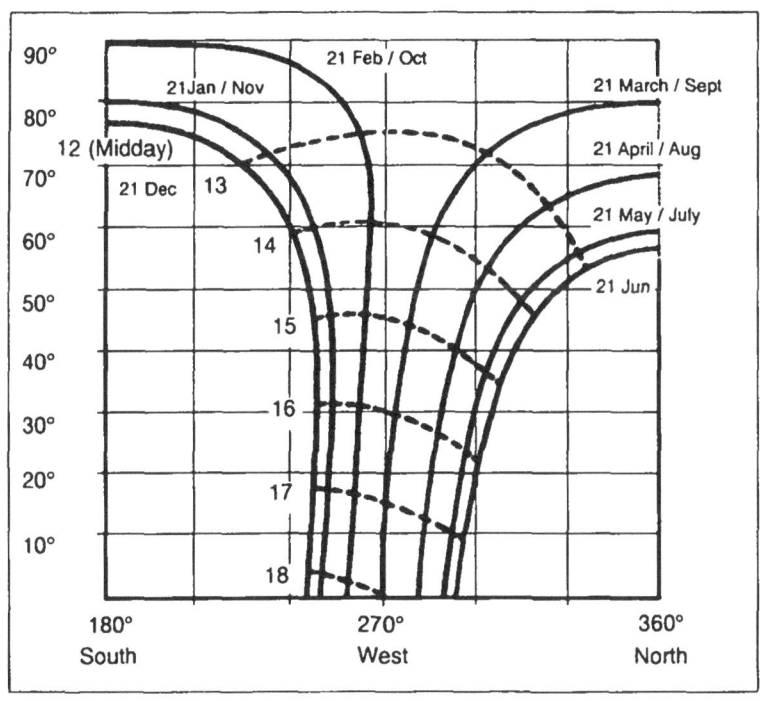

Figure A10 Solar diagram for latitude 10° South

149

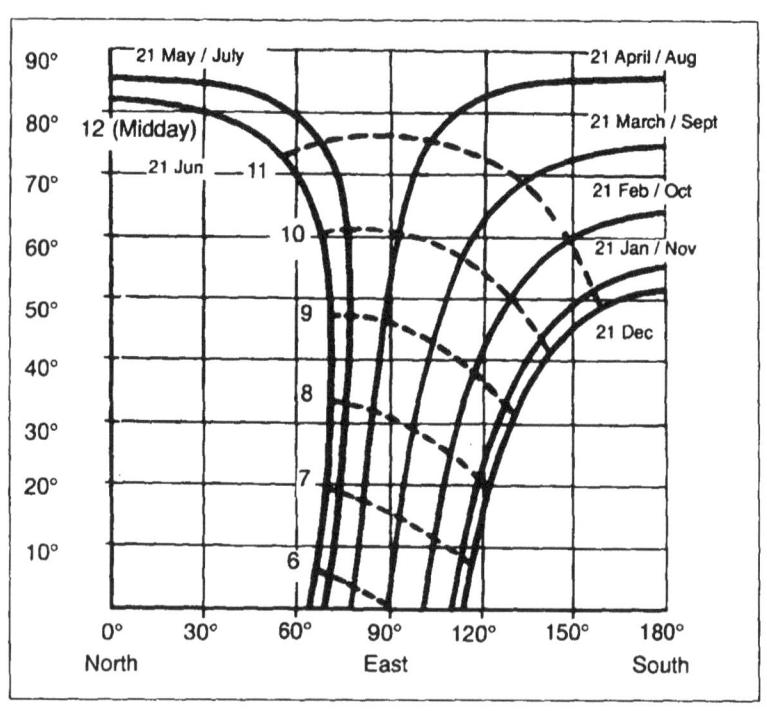

Figure A11 Solar diagram for latitude 15° North

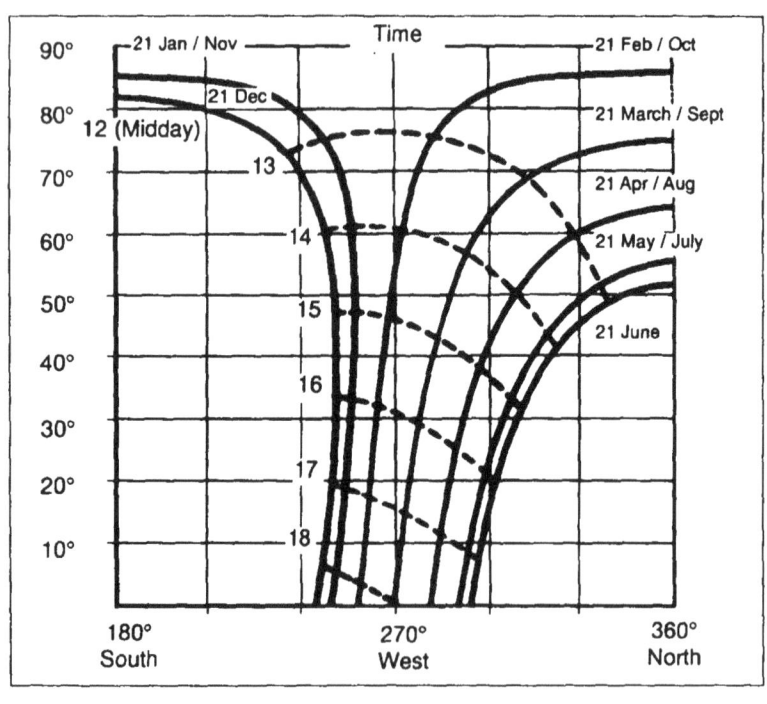

Figure A12 Solar diagram for latitude 15° South

150

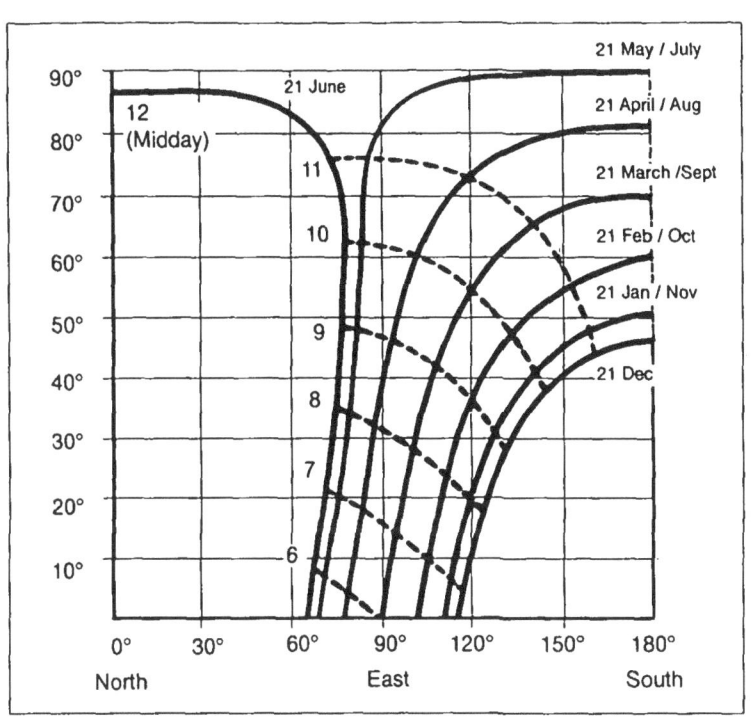

Figure A13 Solar diagram for latitude 20° North

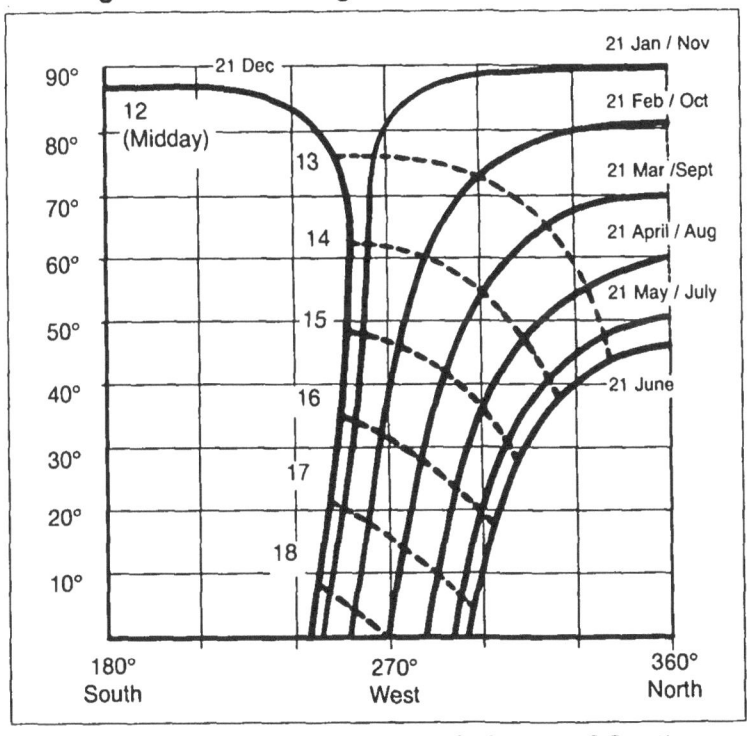

Figure A14 Solar diagram for latitude 20° South

151

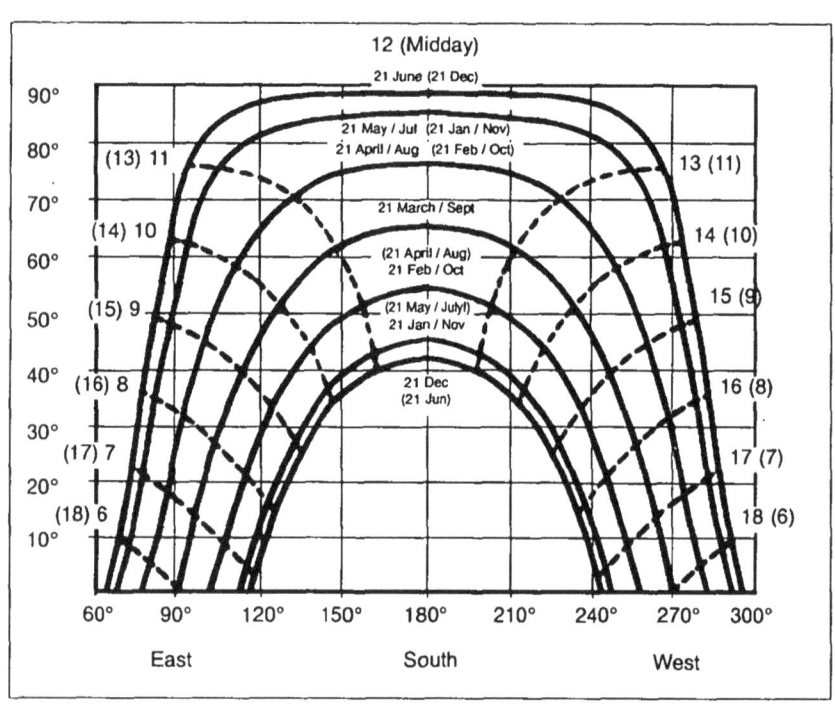

Figure A15 Solar diagram for latitude 25° North

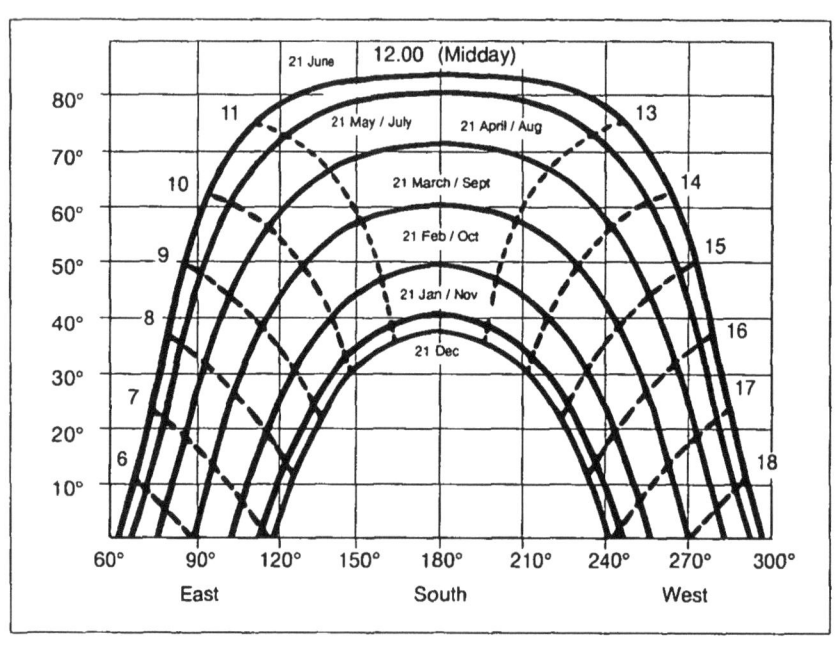

Figure A16 Solar diagram for latitude 30° North

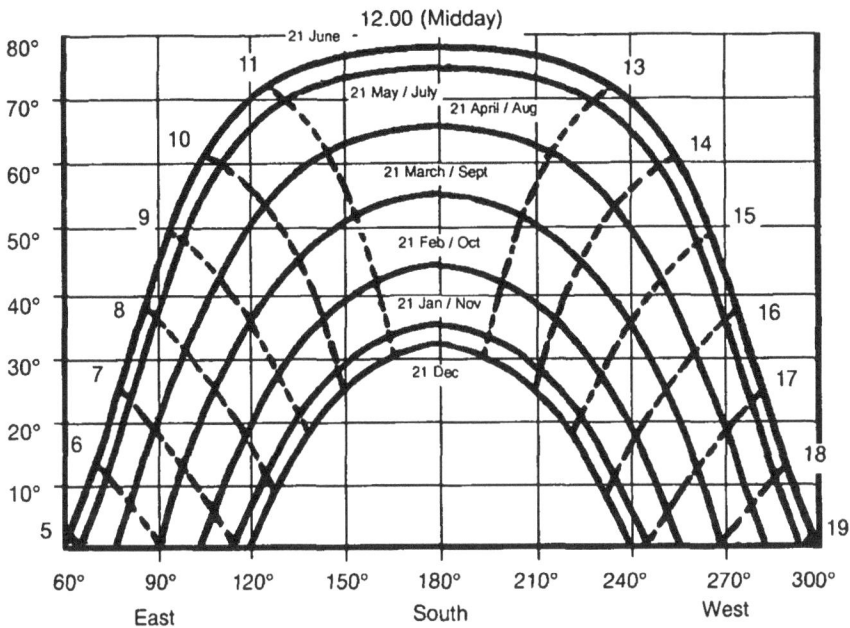

Figure A17 Solar diagram for latitude 35° North

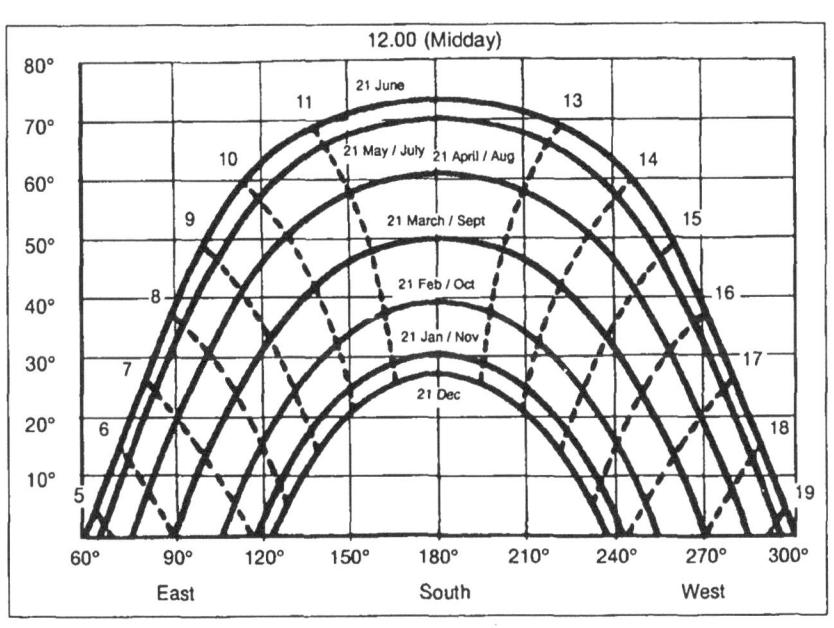

Figure A18 Solar diagram for latitude 40° North

153

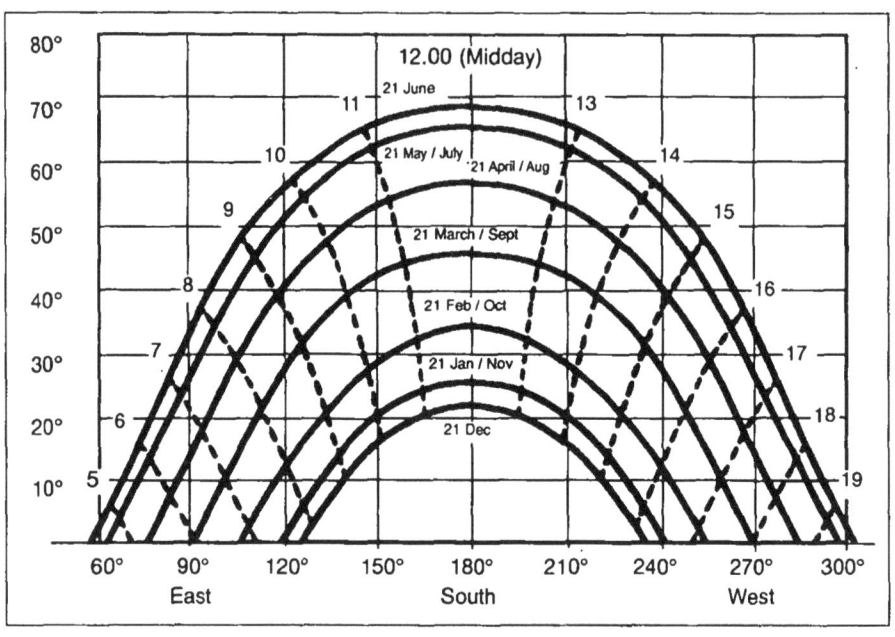

Figure A19 Solar diagram for latitude 45° North

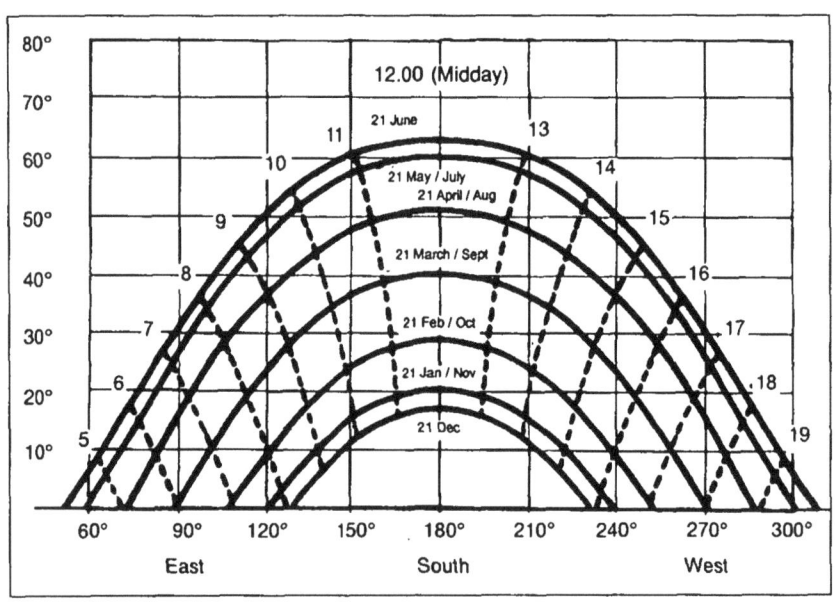

Figure A20 Solar diagram for latitude 50° North

154

Economic calculation and comparative costs

The lifetime cost of a product is defined as its capital cost plus maintenance and operation costs. Thus, in the case of a house the cost of traditional heating using commercial energy can usefully be compared with using solar heating at zero cost (apart from maintenance).

Nevertheless the solar solution implies higher capital cost in all cases: the cost of the solar wall, glazing, etc. It is thus of interest to know when the money saved by using the solar solution is equal to the extra initial investment cost. This time is called the payback period. After that time savings start to be made.

Simplified calculation

Capital cost A = capital cost B   +   additional cost of solar
(solar solution) (traditional solution)

Savings made per year = Operating cost B   –   operating cost A
                        (heating/maintenance) (maintenance)

That is:

Payback period = additional cost of solar ÷ savings made per year

---

Example

There are two neighbours: one chooses to invest in a solar wall, the other decides to continue using occasional heating by means of a stove.

The cost of construction of a solar house is FF4500 more than a traditional house.

Now, each year, with an identical upkeep cost, the solar house can help to save FF1500 compared with the cost of traditional heating (cost of fuel).

The payback time = 4500/1500 = three years

After three years the solar solution has paid for itself and its owner is then making savings on the cost of traditional heating.

Calculation taking into account (a) financing costs (of borrowing) and (b) possible inflation

The above calculation does not apply in the majority of cases because the owner often has to borrow money to finance the additional solar cost. He therefore has to pay interest (financing costs), or the cost of upkeep and heating, which increase each year because of inflation.

(a) Taking account of financing costs

A loan is characterized by its rate of annual interest '*a*' (in %). For a sum 'S' borrowed, $(1 + a).S$ must be paid back after one year, $(1 + a) [(1 + a).S]$ after two years, and so on.

In economic terms, to evaluate the real value of a lump sum, taking account of financing charges (interest rates) the concept of present value accounting is used.

This notion is used to evaluate the savings made by borrowing at the level of interest *a*.

The savings are:

After one year $$\frac{E}{1 + a}$$

After two years $$\frac{E}{(1 + a)^2}$$

Between $n-1$ and n years $$\frac{E}{(1 + a)^n}$$

Discounted savings are less than E, calculated above because the loan must be repaid in order to pay the additional solar cost, increasingly high repayments according to year of repayment, and thus increasingly small savings.

Example
To use the above example again, a neighbour borrows money to finance the additional solar cost at a rate of interest *a* of 10 per cent.

Table A1 Savings allowing for financing costs

Year	Saving on fuel costs	Discounted savings, taking account of loan costs		Total outlay
0				−4500
1	1500	$\frac{1500}{1 + 0.1}$	= 1364	−3136
2	1500	$\frac{1500}{(1 + 0.1)2}$	= 1240	−1896
3	1500	$\frac{1500}{(1 + 0.1)3}$	= 1127	− 769
4	1500	$\frac{1500}{(1 + 0.1)4}$	= 1024	+85

It is now necessary to wait four years before making any real savings over the traditional system. The payback time here is three years and nine months.

(b) Taking account of price variations

The price of maintenance, fuel etc. is subject to inflation of i per cent per year. Savings made must be calculated in real terms. After one year, price P is revalued at $(1+i).P$ and at the end of the second year at $(1+i).(1+i).P$ and so on.

Taking the first case again, with an inflation rate of $i = 2$ per cent.

Table A2 Savings allowing for inflation

Year	Savings (prices constant)	Savings (at 2% inflation)
1	1500	$1500 (1 + 0.02) = 1530$
2	1500	$1500 (1 + 0.02)^2 = 1561$
3	1500	$1500 (1 + 0.02)^3 = 1592$
4	1500	$1500 (1 + 0.02)^4 = 1624$

If inflation and interest rates are taken into account the following is achieved:

Table A3 Taking account of inflation and finance costs

Year	Saving with inflation $(i = 2\%)$	Saving with finance costs $(A = 10\%)$	Discount factor	Total outlay
0				−4500
1	1530	1391	0.909	−3109
2	1561	1289	0.826	−1820
3	1592	1196	0.751	− 624
4	1624	1109	0.683	+ 485

After about three years and seven months of operation, the owner of a solar heated house starts to make savings over traditional house heating, which is less profitable because of increasing fuel costs. Where inflation and interest rates are the same, the two effects cancel each other out and the simplified calculation shown on page 155 then applies.

APPENDIX V

Sizing a *hammam* hearth

Combustion system

Combustion chamber
The volume of the combustion chamber depends on the output of the hearth.

Calculating the required output

This comes back to calculating top-up heating (see method in chapter 10 – stage 5). Once the quantity $Q_{\text{top-up heating}}$ has been determined to achieve the required temperature, the efficiency of a well-designed *hammam* system is estimated to be about 30 per cent.

That is: $Q_{\text{hearth}} = \dfrac{Q_{\text{top-up}}}{0.3}$ in kWh.

The daily combustion duration is determined in hours to provide the required power:

$$P_{\text{hearth}} = \frac{Q\ \text{hearth}}{\text{number of hours combustion}} \text{ in kWh}$$

The volume V is used, where $V = 1\ \text{dm}^3/\text{kW}$

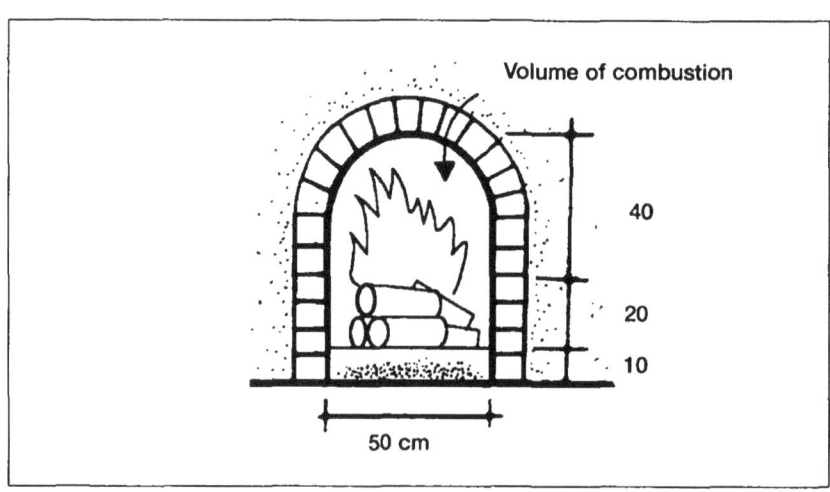

Figure A21 Volume of combustion

If a grate is used the area of the grate is determined so as to provide 40W/cm^2.

In this case, the nominal power is about 60kW (in four hours 50kg of wood are consumed, on the basis that 1kg of wood provides 18 000kJ). With a grate of 1500cm^2 a height of 40cm above the wood is required.

Chimney characteristics

The volume of flue gases to be removed must first be determined. Under normal combustion conditions use a smoke volume equal to 7m^3/kg of wood burnt.

Once the quantity of wood consumed per hour is known (according to the required power level), then the exhaust gas output (D) is known.

The following is a simplified expression (ignoring heat losses under the slab).

$$D = S_{chimney} \sqrt{\frac{2(T_{ch} - T_{ext}).9.81.h}{T_{ext}}} \sqrt{\frac{1}{1 + [\frac{Ss}{Se}]2}}$$

H = 3.5m according to the construction

S_e = 0.3 × 0.3 = 0.09m^2

T_{ext} = 268°K (–5°C, K = 273 +°C)

T_{ch} = 313°K (average value for *hammam* systems)

D = 87m^3/h = 0.024m^3/s

The cross-sectional area of the chimney is hence 60cm^2, that is a diameter of 10cm for a circular cross-section.

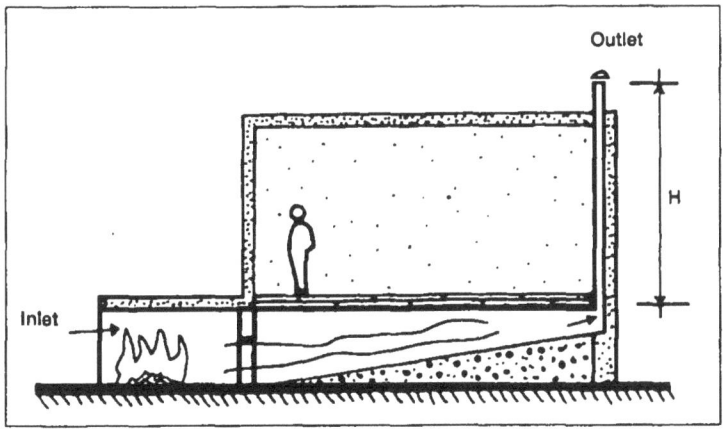

Figure A22 Chimney design

159

Calculating sizes for a latent heat collector

Calculating losses (Q lost)

Assuming that losses are essentially concentrated at roof level, and taking K as the heat exchange coefficient between exterior and interior:

$$Q_{losses} = K_T. \; S_{roof}.(T_{interior}.T_{exterior}) \text{ in W}$$

with

Q_{losses} = heat lost through roof

K_T = heat exchange coefficient through the roof ($6W/m^2/°C$ in this case, $2.5W/m^2/°C$ with an insulated false ceiling).

S_{roof} = area of the roof (m^2)

A distinction is made between :
night time (16 hours in winter season)

$$E_{lost} \text{ night} = K_T.S_T.(T_{int} - T_{.N}). \; 16 \text{ hours (in Wh)}$$
and   *day time* (8 hours)

$$E_{lost} \text{ day} = K_T.S_T(T_{int} - T_{.D}). \; 8 \text{ hours (in Wh)}$$

(for cloudy time only, otherwise use E_{lost} day = 0 Wh)
where:

E_{lost} Energy lost (in Wh)

T_{int} Average interior temperature (°C) 22°C in this case

$T_{.D}$ Average exterior temperature during the day

$T_{.N}$ Average exterior temperature during the night

$$E_{lost} \text{ over 24 hours} = E_{lost} \text{ night} + E_{lost} \text{ day}$$

Calculation of storage volume, V

Assuming that 40 per cent of incident energy is stored in paraffin (a characteristic of the collector system), then

$$V.L_f = 0.4 \; E_{incident}$$

L_f: latent heat of fusion of paraffin

$E_{incident}$: Solar energy arriving at the glazed surface of the collector

Calculation of collector area

The aim is to balance energy lost and energy replaced from storage. Thus

$$A \text{ collector} = \frac{E \text{ lost over 24 hours per } m^2 . A_{roof}}{0.5(V . L_f)}$$

0.5 (50 per cent) efficiency in decreasing storage if the greenhouse containing the paraffin elements is well insulated at night.

Unit conversion tables

To avoid errors and complications, use of the international system (SI) of units is recommended. This is based on the following four units:

o metre (m)	unit of length
o kilogram (kg)	unit of mass
o second (s)	unit of time
o ampere (A)	unit of electric current

Thus, from the laws of physics the following units are obtained:

o newton (N)	unit of force
o newton per square metre or bar (N/m^2)	unit of pressure
o joule (J)	unit of energy
o watt (W)	unit of power

The joule (J) and even the kilojoule (kJ) are very small units and consequently the kilowatt-hour (kWh) is preferred to designate a quantity of energy.

with 1kWh = 3600kJ we give the following corresponding figures:

1 kilocalorie (kcal) = 4.18kJ

and   1 kilocalorie/hour (kcal/h) = 1.16W

o 1 therm (th)   = 100 kilocalories = 1.16kWh
o 1 horse power = 736W
o 1 foot (')   = 30.48cm
o 1 inch (")   = 2.54cm
o 1 BTU*   = 1054.35 joules
o 1 BTU*/h   = 0.2928W
o °F   = 9/5°C +32
o °C   = 5/9(°F − 32)

---

* BTU : British thermal unit

Glossary

ADOBE BLOCKS: technique for preparing soil blocks which consists of forming earth bricks by compacting them and leaving them to dry in the sun.

ANGLE OF INCIDENCE: angle formed by incident ray and the normal (perpendicular) to the receiving surface.

ANGLE OF INCLINATION: angle of inclination relative to the horizon.

ANGLE OF VISION: the angle from which the object is seen. For radiant heat transfer, this angle determines the quantity of energy transmitted.

AZIMUTH OF THE SUN: the angle defined between the horizontal plane and the direction of the sun and the reference direction (full south for the northern hemisphere).

BIOCLIMATIC: a term used to characterize a type of architecture suited to the outside climate, which allows a comfortable interior environment to be created.

DENSITY: ratio of the mass of a certain volume of a body to the mass of an equivalent volume of water (or of air in the case of gases).

EFFICIENCY: a concept from physics expressing the ratio between the energy supplied to a system and the energy actually used (useful energy).

FUSION: transition of an object from solid to liquid state under the action of heat.

HAMMAM: System of water heating and interior heating used in traditional baths (a traditional Islamic design).

INERTIA: in the context of heat it consists of the total masses which can store energy (opaque walls, containers of water etc.).

ISOTHERM: line joining points of the same temperature.

LATENT HEAT: heat released or absorbed on the change of state (solid to liquid for example) taking place at a constant temperature.

MICROCLIMATE: an area incorporated within a large climatic area but possessing its own distinct characteristics. Microclimates can be due to the relief (in mountainous areas), to a large water surface (such as a coastal area) or to large-scale urbanization.

PHOTOSYNTHESIS: the synthesis of organic substances (glucides) from the action of light energy (from the sun or artificial) by chlorophyllic plants.

PHOTOVOLTAICS: the process of transforming electromagnetic energy (from the sun or artificial) into electric current.

PISÉ: technique for constructing earth walls which consists of compacting the earth between planks to the desired thickness.

REFRACTORY CLAY: a clay is said to be refractory when after first firing it is resistant to very high temperatures without undergoing any deformation (little shrinking). It is recommended for stove linings and other combustion chambers.

RELIEF: the uneven features of the earth's surface (in a mountainous region).

SKYLIGHT: the glazed parts placed along the whole length of a building, normally in the upper part of the building at roof level.

SOFTWARE PROGRAM: used in information technology to describe all necessary instructions, written in a language that can be understood by the computer (e.g. basic, fortran, pascal).

THERMOSIPHON: the phenomenon of the movement within a fluid brought about by the temperature difference between two parts of the fluid.

Bibliography

Solar dwellings/solar energy

Title	Author/publisher	Comments
The Passive Solar Energy Book 1979; French edition 1981	Edward Mazria, Rodale Inc., Emmaus, USA	Reference book, very informative on dimensions for passive solar heating processes.
Expansion of the SLR method (Passive Solar Journal) – Vol.1, No. 2 p.67–90 'Heat storage and distribution inside passive solar buildings', L.A. 9694 – RS)	J.D.Balcomb, McFarland, G.F.Jones, Los Alamos National Laboratory, New Mexico, USA	Simplified methods developed by LASL for predicting how well heating needs will be met and the indoor temperature variation (SLR method and DHC).
Méthode B – SOL, 180 pages	Robert Célaire, Pierre Diaz-Pedregal, Jean-Louis Izard	Implementing French rules of SLR method
Solar Greenhouse 1976; French edition 1982	Bill Yanda, Rich Fisher John Muir Publications, New Mexico, USA	Numerous examples of greenhouses and practical recommendations for growing. Simple sizing method.
Solar architecture and earth construction in the northwest Himalaya, 1991, 132 pages	Sanjay Prakash, Development Alternatives, Vikas Publishing Pvt., 576, Masjid Road, Jangpura, New Delhi 110 014	Theoretical study of possible improvements in heating and architecture of housing in Ladakh.
Le gisement solaire, 222 pages	C. Perrin de Brichambaut, C. Vauge, Lavoisier, 11, rue Lavoisier, 75008, Paris	Scientific data for calculation of insolation on the earth's surface

165

'Bis de feu et sociétés rurales: Haut Atlas et région présaharienne (Maroc)', 1991	Laurent Auclair, ENSA, Montpellier	Thesis (French language)
Manual de energia solar, 1986, 68 pages	Frédéric Michaux, Association Runamaqui, 8, rue Bezout, 75014, Paris	Teaching manual for dissemination and use of solar energy (Spanish language)

Case studies

'Soleil du Sud', 1992, 190 pages	Christine Bénard, Dominique Gobin (Runamaqui), Fondation pour le progrès de l'homme (FPH), 38, rue Saint Sabin, 75011 Paris	Critical analysis of Runamaqui's actions concerning a solar poultry programme in Peru (Anccopaccha)
'Rural energy planning in Sikkim' 1991, 259 pages	K. Sudhakar, PPS Gusain, Development Alternatives, Vikas Publishing Pvt Ltd.	Prospective study of energy issues in Sikkim (Indian State) – planning of project activities
A report on the provision of smokeless chulla stoves, 1993	Leh nutrition project, Leh, 194 101, Ladakh, India	Study of health impact on the population following the diffusion of improved stoves
'Trombe walls in Ladakh'	Ladakh Ecological Development Group – Leh, Ladakh, India	Experimental results on the behaviour of buildings equipped with Trombe walls in Ladakh, India

Architecture

'Traité de construction en terre', 1989, 355 pages	CRATERRE, Editions Parenthèses, France	Presentation of the various techniques for earth construction (clay mortars, pisé, adobe blocks etc.) for construction

Agriculture

'Crianza de Pollos'	Runamaqui, 8, rue Bézout, 75014, Paris	Agricultural guide for rural chicken farming

GERES
A development organization

GERES (Groupe Energies Renouvables et Environnement) is a private non-profit-making organization, bringing together researchers, industrialists and developers on the subject of energy and the environment, providing advice, technical assistance, new ideas and project implementation.

In the period of development of renewable energies in France, GERES, established in 1976, was the first organization in the south-west region to promote a 'rational solar energy development'. GERES later extended its field of activities into energy management, energy saving and the environment.

Taking account of the environment and socio-economic context

As an NGO (under French law of associations of 1901) GERES works in developing countries (India, Africa, Central America) to evaluate, identify and provide technical and financial implementation for development projects and sometimes project management, or most often to ensure that projects are implemented through technical assistance by supervisors.

GERES favours the innovative approach to economic development adopted by European professionals, particularly from south-west France (research centre, SME) which includes promoting activities and programmes to export knowledge and products to countries of the Mediterranean basin (Southern Europe and the Maghreb).

As an organization for promoting energy and environmental management, GERES runs a network of expertise and provides information and training. It produces numerous publications.

Flexible operation

With access to an international network of contacts, GERES uses its own skills or those of associate experts; its team of engineers plays the part of broker between the identification of needs and of new markets (socio-economic aspects) and the development of appropriate products (technical aspects).

GERES puts together large economic programmes whose principal technical and financial partners are: the Agence de l'environnement et de la maîtrise de l'energie (Ademe), the ministries in France and the other countries involved, local groups, the European Commission, international organizations, NGOs, professional organizations, universities.

Diverse skills

GERES has acquired and developed skills covering most subjects concerning management of energy and the environment, especially in the following areas:

○ domestic and service sectors: heating, hot water, air conditioning, cooking, food preservation, desalination, lighting.
○ agriculture and livestock: water lifting, greenhouses, food drying and processing, recycling of wastes etc.
○ craft work and industry: energy saving, industrial heat, waste treatment.
○ transport: gas fuel.
○ rural electrification: lighting, radio and television etc., pumping.
○ health: vaccine conservation (cold chain), sterilization.
○ development: socio-economic needs studies, market research, product development, technology transfer, organizing co-operation projects.

GERES

73 avenue Corot – 13013 Marseille (France)

Tel:(33) 91 70 92 93. *Fax:*(33) 91 06 19 46. *E-mail:* geres-fr@worldnet.net

---

www.ingramcontent.com/pod-product-compliance
Ingram Content Group UK Ltd.
Pitfield, Milton Keynes, MK11 3LW, UK
UKHW021830130526
5758IPUK00005B/97